Geochemical Techniques for Identifying Sources of Ground-Water Salinization

■

Bernd C. Richter
Charles W. Kreitler

Bert E. Bledsoe
EPA Project Officer

C. K. SMOLEY

Library of Congress Cataloging-in-Publication Data

Catalog record is available from the Library of Congress.

ISBN 1-56670-000-0

© 1993 by C. K. SMOLEY

Direct all inquiries to CRC Press, Inc.
2000 Corporate Blvd., N.W.
Boca Raton, FL 33431

PRINTED IN THE UNITED STATES OF AMERICA
1 2 3 4 5 6 7 8 9 0

Printed on acid-free paper

ACKNOWLEDGMENTS

Funding for this project was provided by the U. S. Environmental Protection Agency, Robert S. Kerr Environmental Research Laboratory, Office of Research and Development, under Cooperative Agreement No. CR-815748 to the Bureau of Economic Geology, the University of Texas at Austin. Burt E. Bledsoe was EPA project officer. His continued support is appreciated. We thank the reviewers from the U. S. Environmental Protection Agency for their manuscript reviews. Although the information in this document has been funded wholly by the U. S. Environmental Protection Agency, it does not necessarily reflect the views of the Agency and no official endorsement should be inferred.

Manuscript preparation was completed at the Bureau of Economic Geology. Figures were drafted by Maria Saenz, Tari Weaver, Kerza Prewitt, Jana Robinson, Michele LaHaye, Margret Koening, and Joel Lardon under the direction of Richard Dilon. Word processing and typesetting were done by Susan Llyod under the direction of Susann Doenges. The publication was edited by Kitty Chalstromn and designed by Margret Evans. The efforts of all these people are greatly appreciated.

Bernd Richter is presently traveling, driving his VW bus across the highways of America. Any comments or questions can be directed to Mr. Richter at P.O. Box 5543, Austin, TX 78763. Charles Kreitler is a professor in the Department of Hydrology and Water Resources, University of Arizona, where he teaches hydrogeology.

EXECUTIVE SUMMARY

This report deals with salt-water sources that commonly mix and deteriorate fresh ground water. It reviews characteristics of salt-water sources and geochemical techniques that can be used to identify these sources after mixing has occurred.

The report is designed to assist investigators of salt-water problems in a step-by-step fashion. Seven major sources of salt water are distinguished: (1) Natural saline ground water, (2) Halite solution, (3) Sea-water intrusion, (4) Oil-and gas-field brines, (5) Agricultural effluents, (6) Saline seep, and (7) Road salting. The geographic distribution of these sources was mapped individually and together, illustrating which ones are potential sources at any given area in the United States. In separate sections, each potential source is then discussed in detail regarding physical and chemical characteristics, examples of known techniques for identification of mixtures between fresh water and that source, and known occurrences by state. Individual geochemical parameters that are used within these techniques are presented in a separate section, followed by a discussion concerning where and how to obtain them. Also provided is a description of basic graphical and statistical methods that are used frequently in salt-water studies. An extensive list of references for further study concludes this report.

TABLE OF CONTENTS

FIGURES

ix

x

xi

TABLES

1. INTRODUCTION

1.1. Purpose and Use of this Report

The purpose of this report is to summarize geochemical techniques that can be used in studies of salinization of fresh water. The report is designed to assist investigators through detailed discussion of potentially useful chemical parameters and techniques, as well as of physical and geographical characteristics of potential salinization sources.

The topic of salt-water contamination has been extensively researched, as evidenced by the long list of references compiled for this report. No compendium of the overall topic, however, has previously been compiled. The purpose of this document is not to develop new geochemical techniques for identifying sources of ground-water salinity, but to summarize known approaches for all different sources into a single document so that a researcher will have a reference manual that reviews available work.

Salinization of fresh water is perhaps the most widespread threat to ground-water resources. This document deals with geochemical characteristics of the major known sources of salinity, and as such will be helpful to investigators of salt-water problems. The extent to which this document will be of help will depend to a large degree on the background knowledge of the problem and of the investigator. To an experienced researcher in the field of ground-water quality, this document may serve as a summary of and reference to some of the known techniques that are being used. To investigators new in this field, we suggest the following possible methodology of investigation in combination with this report.

Step 1: The general geographic distribution of major potential salinization sources, that is (1) natural saline ground water, (2) halite solution, (3) sea-water intrusion, (4) oil-and gas-field brines, (5) agricultural effluents, (6) saline seep, and (7) road salt, is addressed in Chapter 2 of this report. Through a series of maps that show the distribution of each source as well as the overlap between these sources, the investigator can get a general idea which potential salinization source or sources exist at her/his local area of interest at any given area of the country. There are some limitations to the maps, as discussed in Chapter 2.

Step 2: After potential sources of salt water have been identified, Chapter 3 should be consulted for a discussion of the sources. This will provide the researcher with the necessary background information about the source(s) of interest. Each of the seven sources is discussed in detail, including mechanisms of mixing with fresh ground water, chemical characteristics, geochemical case studies, recommended chemical techniques for identification of salinization caused by these sources, and a state-by-state summary of occurrences. For each source, a variety of techniques that can be used is presented.

1

Each section includes a state-by-state summary of known problem cases associated with individual salinization sources. Before disregarding any source identified in Step 2 as absent in the area of salinization and of interest, the investigator may want to review all summaries pertinent to her/his state(s). With the help of references listed in Chapter 7, extensive background information of the problem can be obtained.

Step 3: After selecting techniques that are useful for the particular problem case, the geochemical parameters of interest should be reviewed in Chapter 4. This will give the investigator a general overview of parameter characteristics as well as sampling techniques and costs of laboratory analyses.

Step 4: Depending on the area of interest, chemical data may or may not be available to the investigator from published sources, agency files, or computerized data banks. Some of the techniques selected in Step 3 may be applicable using existing data, but others most likely will necessitate collection of water samples for parameters that are not determined on a regular basis (for example, isotopes). Chapter 5 should be consulted for guidelines of data selection and for a discussion on computerized data banks. This step is crucial, as existing data can be very helpful but may also be misleading. Chemical analyses that may be representative of potential salinization sources can be found in the referenced literature.

Step 5: Once data have been selected from existing sources or collected in the field, evaluation can be accomplished using techniques selected during Step 2. Useful graphical and statistical methods are discussed briefly in Chapter 6. Hopefully, the source of salinity will then be determined.

1.2. Background

All natural waters contain some dissolved minerals through the interaction with atmospheric and soil gases, mixing with other solutions, and interaction with the biosphere and lithosphere. In many cases, these processes result in natural waters that contain total dissolved solids (TDS) concentrations above those recommended for drinking water (Table 1). This deterioration of water quality is enhanced by almost all human activities through water consumption and contamination.

Salinization, that is, the increase in TDS, is the most widespread form of water contamination. The effect of salinization is an increase in concentrations of specific chemical constituents as well as in overall chemical content. A variety of terms have been introduced in the literature to reflect the changing character of the water as salinity increases, such as saline, moderately saline, very saline, brackish, and brine (Table 2). For the purpose of this report, we followed the classification of Robinove and others (1958), which is one of the most widely used. The term "salinization," as it will be used in this report,

Table 1. Drinking-water standards established for
inorganic consituents (data from Freeze and Cherry,
1979, and U.S. Environmental Protection Agency, 1989).

Constituent	Recommended concentration limit (mg/L)
Total Dissolved Solids (TDS)	500.000
Chloride (Cl)	250.000
Sulfate (SO_4)	250.000
Nitrate (NO_3)	45.000
Iron (Fe)	0.300
Manganese (Mn)	0.050
Copper (Cu)	1.000
Zinc (Zn)	5.000
Boron (B)	1.000
Hydrogen Sulfide (H_2S)	0.050

	Maximum permissible concentration (mg/L)
Arsenic (As)	0.050
Barium (Ba)	1.000
Cadmium (Cd)	0.010
Chromium (Cr)	0.050
Selenium (Se)	0.010
Antimony (Sb)	0.010
Lead (Pb)	0.050
Mercury (Hg)	0.002
Silver (Ag)	0.050
Fluoride (F)	1.4–2.4

Table 2. Ground-water classification based on TDS
ranges (in mg/L).

Robinove and others, 1958:

Fresh	0	–	1,000 TDS
Slightly saline	1,000	–	3,000
Moderately saline	3,000	–	10,000
Very saline	10,000	–	35,000
Briny	>35,000		

Freeze and Cherry, 1979:

Fresh water	0	–	1,000 TDS
Brackish water	1,000	–	10,000
Saline water	10,000	–	100,000
Brine	>100,000		

indicates an increase in TDS from background levels by any source. As such, salinization may or may not cause concentrations higher than drinking water standards.

Of the variety of potential sources of salinity, some are natural and others are anthropogenic. Precipitation interacts with atmospheric gases and particles even before it reaches the earth's surface, as reflected in often low pH values in areas of high sulfur dioxide content in the atmosphere (formation of sulfuric acid, "acid rain"). Strong winds carry mineral matter and solution droplets (for example, ocean spray) that can be dissolved and incorporated into precipitation. Surface runoff dissolves mineral matter on its way toward a surface-water body, where it mixes with water of different chemical composition. Water that enters the soil is subject to additional chemical, physical, and biological changes, such as evapotranspiration, mineral solution and precipitation, solution of gases, and mixing with other solution. Changes in chemical composition continue in ground water along flow paths from recharge areas to discharge areas. Water-rock interaction and mixing are the dominant processes. Mixing of different waters is often enhanced by human activities. For example, improper drilling, completion, and final construction of wells may create artificial connections between fresh-water aquifers and saline-water aquifers. Pumping of fresh water may change directions of ground-water flow and may cause encroachment of saline water toward the pumped well; improper waste-disposal activities or techniques may introduce artificial solutions that contaminate natural ground water.

Some areas of the country experience very little problems regarding salinization of fresh-water resources, whereas in other areas most of the available ground water is saline, reflecting natural and human-induced degradation. Where such conditions for salinization of fresh water exist is discussed in the following chapter.

2. GEOGRAPHIC DISTRIBUTION OF MAJOR SALINIZATION SOURCES

For the purposes of this report, seven major salinization sources were singled out. They are, (1) natural saline ground water, (2) halite solution, (3) sea-water intrusion, (4) oil-and gas-field brines, (5) agricultural effluents, (6) saline seep, and (7) road salt. The geographic distribution of these potential sources and areas of overlap between these sources are discussed in this chapter. A detailed discussion of each individual source will follow in the next chapter.

Saline ground water (TDS>1,000 parts per million [ppm]) of variable origin underlies approximately two-thirds of the United States (Feth and others, 1965). It may be encountered in water wells that were drilled too deep for local conditions and it is a threat to those wells that are pumped at a sufficiently high rate to induce salt-water flow toward the well. Shown in figure 1 are those areas where TDS concentrations are greater than 1,000 ppm within 500 ft of land surface. Outside these areas, saline water does occur, but generally at depths greater than 500 ft below land surface (Feth and others, 1965). Although the potential of salinization exists in these outside areas wherever wells are pumped at high rates and from great depths, the cut-off value of 500 ft was adopted from Feth and others (1965) because the search for usable ground water (and with it drilling activities to greater depths) may be greatest in areas where less fresh water is available. Conditions may change in the future or may be different locally, as the demand for ground water increases.

Many sedimentary basins are known to contain large deposits of rock salt in the form of salt beds or salt domes (Fig. 2). Some of these deposits occur at great depths, such as those in southernmost Florida that are at greater depths than 10,000 ft below land surface. Others occur close to land surface, such as those in parts of Utah (Dunrud and Nevins, 1981). Shallow occurrences of salt in Texas, Louisiana, Alabama, and Mississippi along the Gulf of Mexico are due to salt diapirism. The presence of salt deposits millions of years old indicates a relatively high stability, that is, little contact with ground water. Where ground water comes into contact with salt deposits, often enhanced by heavy drilling and mining activities, especially in salt-dome areas, solution of salt and salinization of local ground waters will occur. The shallower the salt deposit, the higher the potential of fresh-water salinization. In this report, all salt deposits are considered potential salinization sources, regardless of the depth of occurrence.

Where coastal aquifers are interconnected with the open ocean, sea-water intrusion can occur. Where formation water hasn't been flushed out, where sea-water has intruded or is intruding into coastal aquifers as a result of high sea-water levels, or where pumping induces landward flow of sea water, the potential of well-water salinization exists. For the purpose of this report, all coastlines of the country were considered potential intrusion areas, regardless of the nature of the coastal aquifer (Fig. 3). On a local scale, some coastal areas can probably be disregarded as a source of salinity, especially where ground-water pumpage is low.

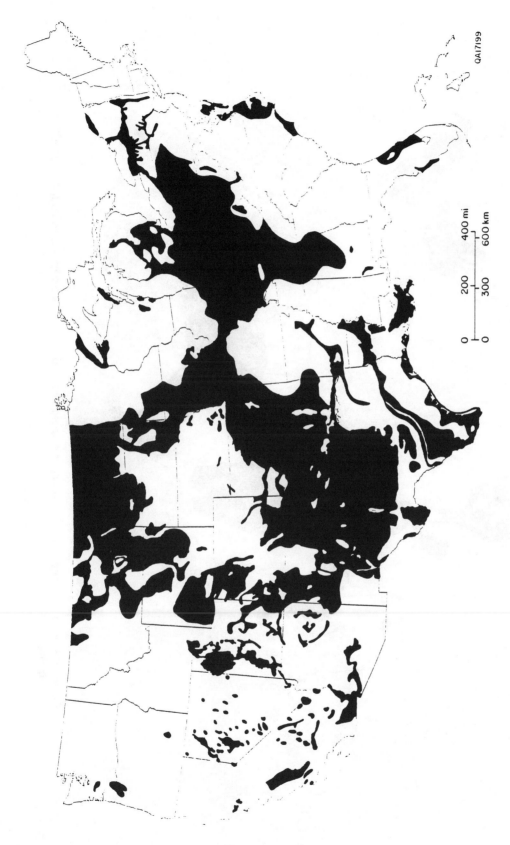

Figure 1. Map of the United States showing areas of ground water containing more than 1,000 mg/L total dissolved solids at depths less than 500 ft below land surface (data from Feth and others, 1965).

QAI7I99

7

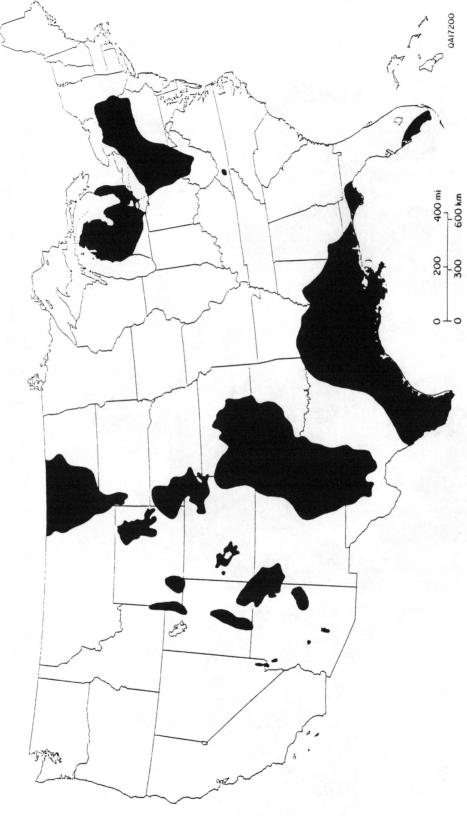

Figure 2. Map of the United States showing the approximate extent of halite deposits (data from Dunrud and Nevins, 1981).

QAl7200

8

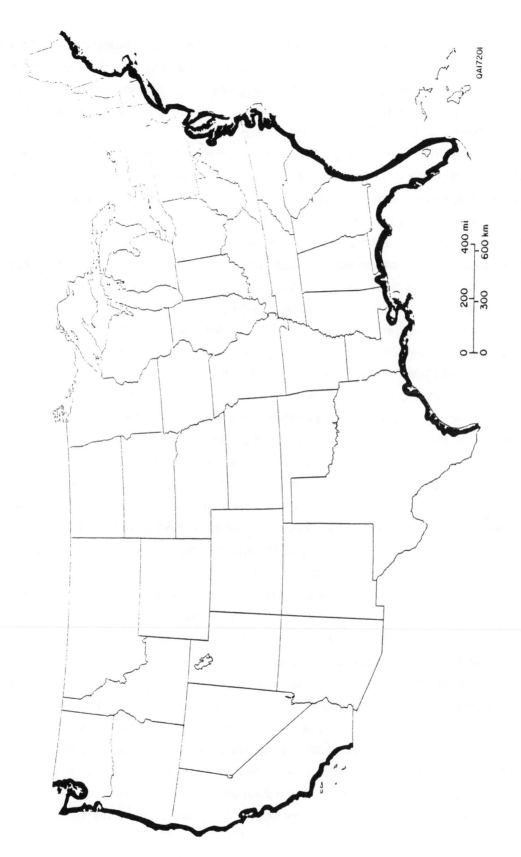

Figure 3. Map of the United States showing areas of potential sea-water intrusion.

QAI7201

200	400 mi	
300	600 km	

9

Associated with the exploration of oil and gas is the creation of avenues for water migration from great depths into the shallow subsurface. Subsequent production brings huge amounts of brine to the land surface. These drilling activities and the disposal of these brines are some of the biggest salinization hazards in the country. Parts of 25 producing states are potentially affected by this hazard, as mapped by the general distribution of oil and gas fields in the United States (Fig. 4).

Salinization as a result of agricultural activities is found all over the country. Irrigation-return waters pose a potential threat in the western half of the United States (Fig. 5), where precipitation rates are low and where evapotranspiration rates and salt contents in soil are high. Another salinization source enhanced by agriculture is dryland saline seep. Terracing of land and destruction of natural vegetation caused this phenomenon in several states, resulting in the salinization of soil and ground water (Fig. 6).

Weather conditions favor concentration of road salting in the northeastern part of the country (Fig. 7). There, millions of tons of salt are applied to roads each winter, imposing a salinization threat to soil, plants, and surface and ground water in the vicinity of highways.

Mapping of potential salinization sources, as done in figures 1 through 7, is helpful in determining sources of salinity at any particular area in the country. By overlaying all seven sources, a variety of combinations between these sources becomes evident. This large variety complicates generic approaches to salt-water studies, because salt-water characteristics change considerably from area to area depending on the kind of combination of sources involved. In addition, not only the potential salinization sources change from area to area, but also the chemical characteristics of individual sources may not be the same everywhere, greatly increasing the number of potential combinations of possible mixing between fresh-water and salt-water sources. As the composite map (Fig. 8) of the above-mentioned potential sources indicates, approximately three-quarters of the country could possibly be affected by two or less than two of the selected sources. In these areas, identification of an actual salinization source could be easier than in other parts, where three or more potential sources exist.

There are several limitations to the applicability of the maps (Figs. 1 through 8) presented here. Limitations that should be kept in mind when selecting or eliminating potential sources at any given area are (1) only the sources discussed in this report were mapped; (2) the sole presence of a potential source does not necessarily indicate that the source actively contributes to any salinity problem; (3) effects of these sources may be felt away from the point of origin; (4) areas of occurrence were generalized with approximated boundaries; and (5) the known distribution of potential salinization sources may have changed since originally mapped. Not included on these maps are some other known salinization sources, such as sewer systems, thermal springs, waste-disposal facilities, or mining areas, all of which may contribute to ground-water salinization in some areas. These sources were not considered because large-scale regional mapping would have been more complicated and would have resulted in even more small areas of overlap. Also, chemical characteristics of these sources vary to a higher degree locally than the ones discussed here, which would have complicated the discussion even further. The seven

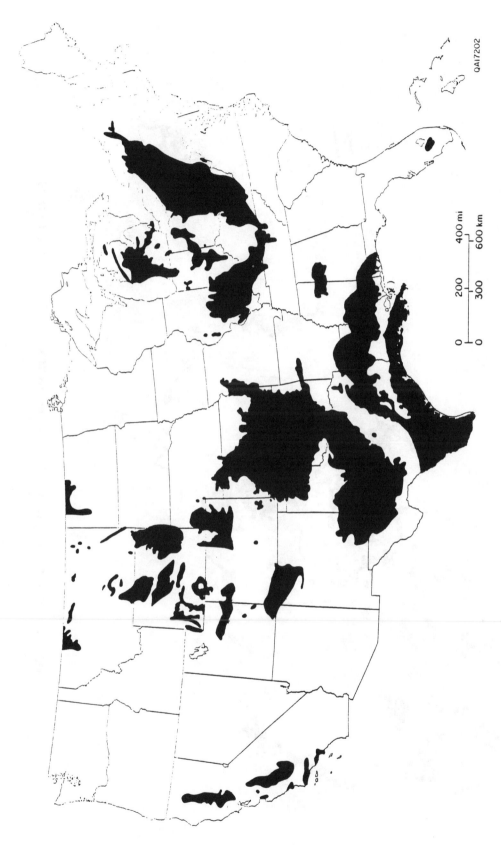

Figure 4. Map of the United States showing general areas of oil and gas production (modified from PennWell Publishing Co., 1982).

QAI7202

11

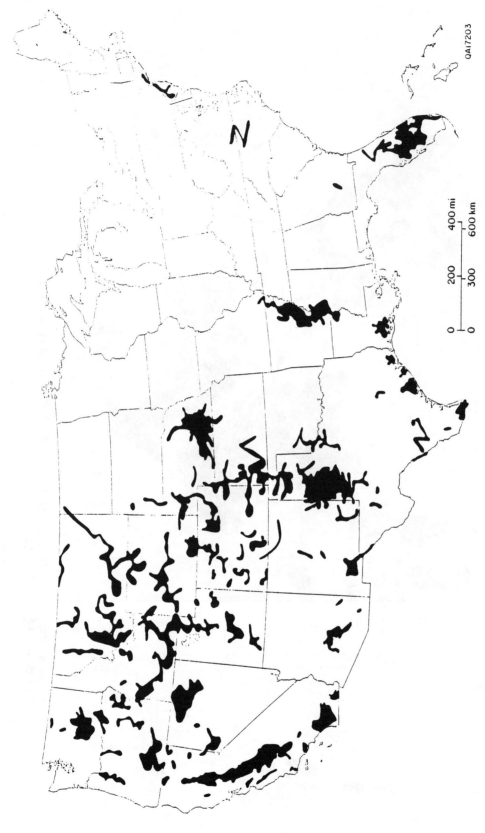

Figure 5. Major irrigation areas in the United States (modified from Geraghty and others, 1973).

QAI7203

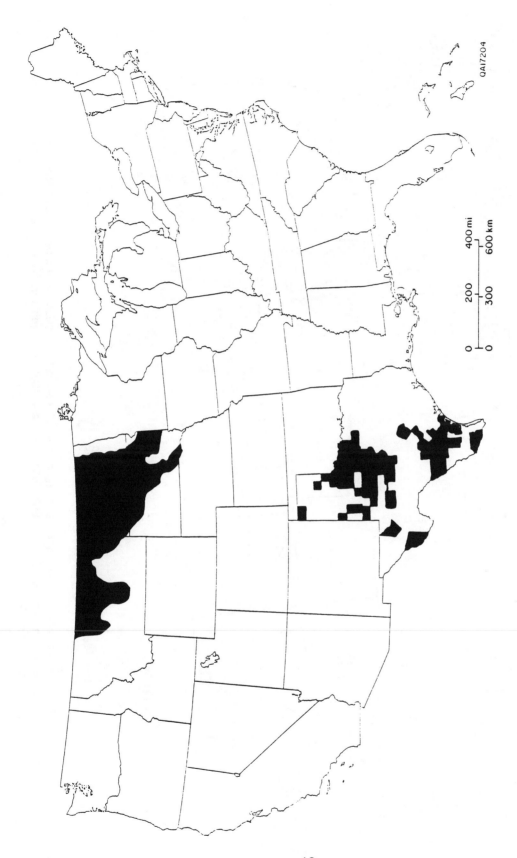

Figure 6. Saline-seep areas in Montana, North and South Dakota, and Texas (data from Bahls and Miller, 1975, and Neffendorf, 1978).

QAI7204

400 mi
600 km
200
300
0
0

13

Figure 7. Map of the United States showing amount of road salt (in thousands of tons) used in individual states during the winters of 1966–1967, 1981–1982, and 1982–1983 (data from Field and others, 1973; Salt Institute, undated).

EXPLANATION

4/85/98   1966-67/1981-82/1982-83

NR   Not reported

14

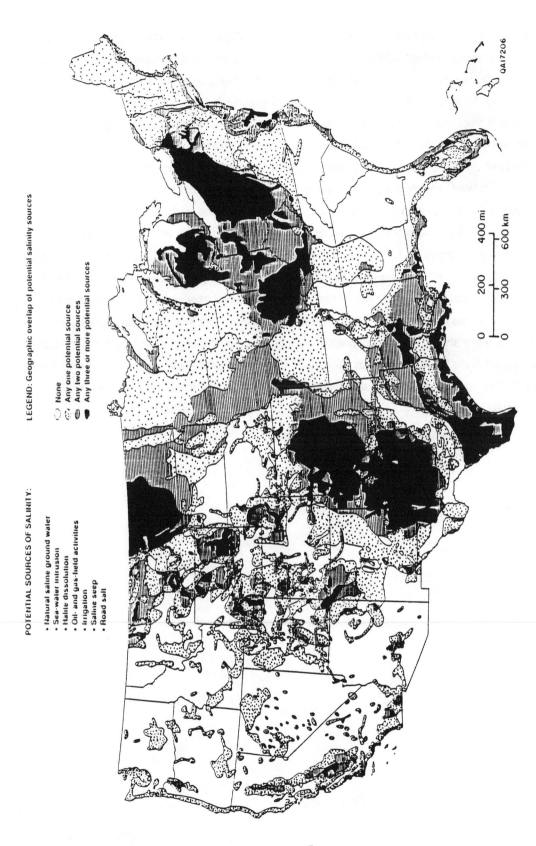

POTENTIAL SOURCES OF SALINITY:

• Natural saline ground water
• Sea-water intrusion
• Halite dissolution
• Oil- and gas-field activities
• Irrigation
• Saline seep
• Road salt

LEGEND: Geographic overlap of potential salinity sources

○ None
•• Any one potential source
≋ Any two potential sources
● Any three or more potential sources

0   200   400 mi
0   300   600 km

QAI7206

Figure 8. Composite map of potential salinization sources as mapped in figures 1 through 7.

15

potential sources mapped may or may not be active at any area, depending on natural and human-induced conditions. For example, oil-field activities or the presence of halite in the subsurface don't necessarily imply that either cause salinity of ground water in a specific area. Also, for reasons of simplicity, only those states with a long history of road-salt use (cut-off value of greater than 30,000 tons of road salt use per year in 1966) was used in the overlap map of figure 8. This should not imply that, for example, a storage area of road salt in any of the states that applied less than 30,000 tons per year at that time is not a potential source of salinity, or that increased usage in some other states may not pose a problem now or in the future. If ground-water salinity is a problem, every potential source in that area should be considered. If a salinity problem exists in an area where no potential salinization source is indicated on figures presented here, either transport of salt water from a bordering area or from a source not covered in this report should be considered. It also should be kept in mind that some of the base maps used in constructing these maps are up to 30 years old, and may be somewhat outdated regarding the present occurrence of any of these sources on a local basis. In any case, these maps should probably be used only to get a general idea of potential sources involved, after which available maps with more local detail should be consulted. The scale factor alone should make it imperative that the geographic distribution of salinization sources should be mapped at the beginning of each salinization study using local maps instead of maps presented in this report. The sources mapped in this chapter will be discussed individually and in detail in the following chapter.

3. MAJOR SALINIZATION SOURCES

As discussed in the introduction chapter, a variety of salinization sources exists throughout the country. Some of these are more dominant than others, and individually sources may be active in some place but not in another, depending on natural conditions and changes induced by man. Individual salinization sources judged as most important on a regional level are (1) naturally occurring saline ground water, (2) halite solution, (3) sea-water intrusion, (4) oil-field and gas-field brines, (5) agricultural by-products and techniques, (6) saline seep, and (7) road salt.

Each section on individual salinization sources is divided into (1) a discussion of the mechanism of contamination, (2) the chemical composition of the salinity source, (3) examples of geochemical studies that were conducted to identify the specific salinity source, (4) the most significant chemical parameters that are commonly used to trace the respective source, and (5) a state-by-state summary of the salinity source. This way, a researcher is provided with an in-depth review of geochemical methods for identifying various salinity sources. A cookbook approach for identifying salinity sources is not followed because the complexity of local contamination and hydrogeology precludes a step-by-step approach. After reviewing the information contained in a section, a researcher needs to develop his/her own hydrochemical criteria based on the type of salinity, the hydrogeologic setting, and the type of data and budget available.

Although comprehensive, the following discussion of salinization sources cannot be complete. Local sources of reference for the area of interest should be incorporated by the researcher, who should have a general understanding of ground-water conditions in the area.

3.1. Natural Saline Ground Water

3.1.1. Mechanism

Natural saline ground water, as used for this manual, is regionally occurring saline ground water that underlies fresh-water aquifers. This chapter deals with natural discharge of such saline ground water, with pumping-induced mixing between saline ground water and fresh water, and the upward migration of saline ground water along boreholes drilled through the fresh-water section into the salt-water section. Not discussed is salinization associated with solution of halite (see chapter 3.2), with sea-water intrusion (see chapter 3.3), and with oil-field brine production (see chapter 3.4), all of which also deal with natural saline ground waters. Chemical characteristics of deep-basinal formation brines are similar or identical to most brines produced from oil and gas reservoirs. The nature of occurrence in salt-water problems is different, however, as contamination by oil-field brines involves pumping and disposal of brine or creation of artificial migration pathways, whereas contamination by deep-basin brines involves subsurface migration as a result of natural or pumping-induced conditions. This difference in mechanisms of contamination but

17

similarity in chemical characteristics creates some overlap in the discussion of natural saline ground water in this section and the discussion of oil-field brines in chapter 3.4.

Saline ground water underlies the country at variable depths. Some areas don't contain any fresh ground water, whereas in other areas thick fresh-water aquifers overly saline ground water. For example, little or no saline water is known to occur within 1,000 ft of land surface in most of Nebraska and Missouri, whereas in some other states, such as Indiana, Ohio, and North Dakota, saline water is encountered at less than 500 ft below land surface throughout almost the entire state (Fig. 1). Where plenty of fresh water overlies shallow saline water, no major problems of salt-water mixing with fresh water may occur, such as in Iowa (Atkinson and others, 1986). Other areas may not be so fortunate, such as New Mexico, where an estimated 75 percent of all ground water is too saline for most uses (Ong, 1988).

The occurrence of saline ground water is dependent on a variety of factors, including distribution and rates of precipitation, evapotranspiration and recharge rates, type of soil and aquifer material, residence time and flow velocities, or nature of discharge areas. The origin of natural saline ground water can be residual (connate) water from the time of deposition in a saline environment, solution of mineral matter in the unsaturated and saturated zones, concentration by evapotranspiration, intrusion of sea water, or any mixture of the above. Residual saline water is not found very often within the shallow subsurface because of the normal flushing of formation water by precipitation through time. Relatively young coastal aquifers may still contain pockets of connate water where hydraulic gradients are low and where hydraulic conductivities are low. Typically, natural salinity in ground water increases with depth below land surface as chemical reactions with aquifer material, resident time, and mixing of different waters increase. Ground water in discharge areas typically is of lower quality than ground water in recharge areas because of water–rock interactions and possible mixing with saline water along the flow path (Fig. 9).

In some settings, especially in the western half of the United States, salts may be concentrated in the shallow subsurface due to evaporation rates that exceed precipitation by up to one order of magnitude (Geraghty and others, 1973). Significant recharge pulses may dissolve this salt and flush it into ground water. Evaporation is enhanced by transpiration of water by plants, which is a serious problem in several southwestern states. Woessner and others (1984) reported that phreatophytes can cause TDS increases from 2,000 mg/L to 11,000 mg/L in a single growing season in parts of Arizona and Nevada. The salt that accumulates in the soil during the growing season, from 403 mg/kg to 28,177 mg/kg, is flushed toward discharge areas during major recharge periods.

Many ground-water basins in the western United States are closed. In those basins, natural recharge along the surrounding highlands flows toward the basin centers. Along the flow path, ground water dissolves mineral matter, resulting in a general increase in TDS content from recharge areas to discharge areas. Evaporation in the basins and especially in salt flats in the center of the basins is the most influential process in the development of the chemical composition of the shallow, saline ground water in these settings (Boyd and Kreitler, 1986). Evaporation and mineral precipitation concentrate the ground water to

18

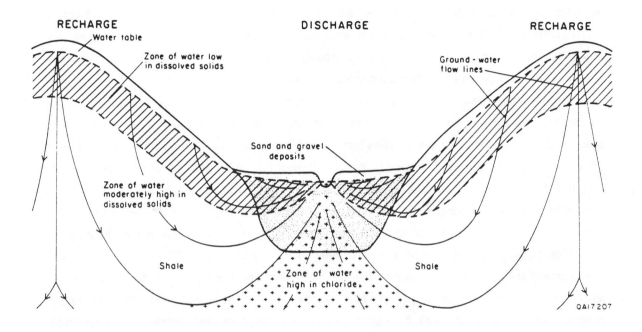

Figure 9. Relationship between ground-water quality and recharge-discharge areas (from Kantrowitz, 1970). Solution of aquifer material and evapotranspiration in discharge areas commonly cause water-quality deterioration along the flow path.

the composition of a Na-Mg-SO$_4$-Cl-type brine, noted for Deep Springs Lake (Jones, 1965), Death Valley (Hunt and others, 1966), and Sierra Nevada Basin (Garrels and MacKenzie, 1967) in California, for Abert Lake in Oregon (Jones and others, 1969), for Teels Marsh in Nevada (Drever, 1982), and for the northern Salt Basin of West Texas (Boyd and Kreitler, 1986). Surface water that accumulates at topographically low areas in the center of the basins is often concentrated by evaporation. Infiltration of these concentrated waters can reach the water table and cause severe pollution. In the western United States, many areas which are now drained by major river systems may have been under closed conditions in the past, during which high concentrations of residual salt were left behind in low-permeable, lacustrine deposits. Because of low permeabilities, these salts may have been preserved in these settings under natural conditions. Overpumpage of good-quality ground water can lead to inflow of these saline ground waters, as fresh-water sources are depleted and waters are drained out of low-permeable units.

Natural contamination of fresh water by saline ground water occurs where salt water from saline aquifers (a) discharges at land surface or (b) mixes with fresh water in the subsurface. (a) Natural saline springs have been reported at many localities in the United States, such as in New York by Crain (1969), in Oklahoma by Ward (1961), in Texas by Richter and Kreitler (1986a,b), or in Arizona by Fuhriman and Barton (1971). Some of these natural springs contain TDS of up to 300,000 mg/L as a result of solution of halite in the shallow subsurface (see also chapter 3.2.). Geothermal springs associated with fault zones or volcanic activity are often mineralized. In other areas, saline-water discharge occurs as nonpoint contamination, such as along the Colorado River, where more than 50 percent of the average annual 10.7 million tons of salt is contributed by diffuse seepage (U.S. Department of Agriculture, 1975; Atkinson and others, 1986). (b) Mixing of saline ground water and fresh recharge water occurs in discharge areas of regional and local aquifer systems, such as in the outcrop areas of Permian formations in north-central Texas (Core Laboratories, Inc., 1972) and of Mississippian formations in central Missouri (Fig. 10) (Banner and others, 1989). Fresh-water and salt-water facies are generally separated by a zone of mixing of variable thickness. The position of this mixing or transition zone may vary in response to changes in either flow component, which, on a nongeological time scale, is mostly in response to heavy pumpage of fresh water. Drawdown of the water table or the potentiometric surface in some aquifer systems is so severe that this interface moves to within the cone of depression of individual wells or well fields, resulting in contamination of wells after some time of pumping. Mixing between fresh water and saline water can also occur through any type of boreholes that penetrate and are open to both types of waters. If heads in the fresh-water unit are higher than in the salt-water unit, fresh-water will drain into the saline formation. If salt-water heads are higher than fresh-water heads, contamination of an entire aquifer in the vicinity of the well may occur.

Intermittent pumping of wells can lead to changes in water quality associated with chemical reactions within the cone of depression. Oxidation of pyrite within the cone during pumping of the well and the subsequent solution of iron and sulfate during the recovery of water levels to prepumping levels when the

20

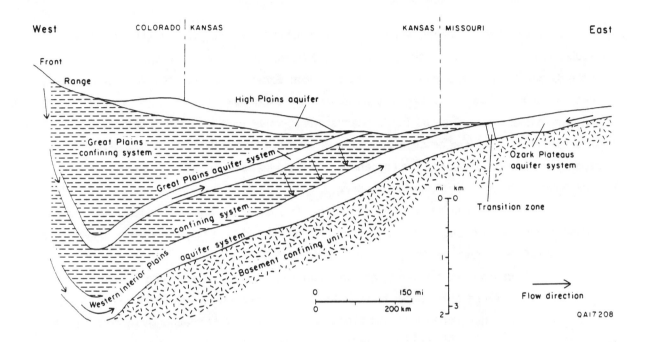

Figure 10. Discharge of regional, saline ground water and mixing with local, fresh ground water. Meteoric recharge in the Front Range of Colorado acquires high salt contents from Permian evaporites under the High Plains aquifer along its flow path to recharge areas in central Missouri, where it mixes with local meteoric, fresh ground water (modified from Banner and others, 1989).

well is not pumped, often leads to high iron and sulfate concentrations in well waters (Custodio, 1987). Changes in water quality during pumping may also depend on the depth of the well. For example, in a salt-water study in Arkansas, Morris and Bush (1986) sampled two adjacent wells of different total well depths (Fig. 11a). Over time, chloride content increased in the shallow well and decreased in the deep well (Fig. 11b). The increase in the shallow well suggested inflow of deep saline water, whereas the decrease in chloride in the deep well suggested inflow of fresh water (Morris and Bush, 1986).

3.1.2. Hydrochemistry of Different Sources of Naturally Occurring Salinity

Nearly all geologic environments may contain naturally-occurring saline water resulting from geochemical processes within each geologic setting. Natural, highly saline waters typically have chloride as the dominant anion and sodium as the dominant cation. Exceptions are waters associated with saline seep and some salt flats, which often have sulfate as major anion. Calcium concentrations are sometimes very high in deep-basin brines, as TDS concentrations approach several hundreds of thousands mg/L. These brines are comprised almost entirely of NaCl and $CaCl_2$. Natural brines associated with mineral deposits often contain unusually high concentrations of ions that are normally not concentrated in most other brines, such as Cu, Zn, Ni, Co, Mb, Pb, or Ag.

The origin of the chemical composition of brines in sedimentary basins is widely discussed and sometimes disputed. Depending on the hydrologic environment of each respective basin, the chemical composition of saline water may differ, as outlined by Kreitler (1989):

Basin waters often are referred to as either connate, meteoric, or a mixture of both. The term "connate" is defined for this paper as water that was trapped at the time of sedimentation. The term "meteoric" indicates ground water that originated as continental precipitation. By definition, the age of connate waters coincides with the age of the host sediments. Basinal waters of meteoric origin are younger than the host sediments. Though most basins are composed predominantly of marine sediments, formation waters generally do not resemble sea water in either chemical composition or concentration. Many basins contain waters with total dissolved solid concentrations significantly greater than sea water concentrations and as high as 400,000 ppm. Maximum salinities in the Tertiary section of the Gulf Coast Basin are approximately 130,000 ppm; in the East Texas Basin, 260,000 ppm; in the Palo Duro Basin, 250,000 ppm; in the Illinois Basin, 200,000 ppm; in the Alberta Basin, 300,000 ppm; and in the Michigan Basin, 400,000 ppm (Bassett and Bentley, 1983; Hanor, 1983). Two different types of brines are generally found, a Na-Cl brine and a Na-Ca-Cl brine, neither having chemical composition ratios similar to sea water.

In recent years, three mechanism have been used to explain the high ionic concentrations and the chemical composition of the brines (Hanor, 1983) (1) the brines originated as residual bittern brine solutions left after the precipitation of evaporites; (2) basinal waters have dissolved halite that was

22

(a)

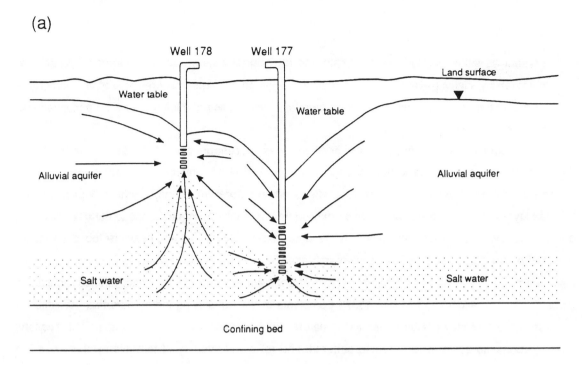

(b)

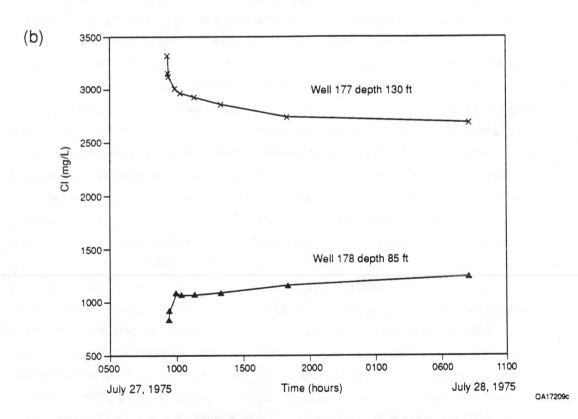

Figure 11. Relationship between depth of screen and chloride content in two adjacent wells (from Morris and Bush, 1986). (a) Pumpage-induced changes in flow directions cause upward migration of saline water into fresh water in well No. 178 and downward migration of fresh water into saline water in well No. 177, resulting in (b) water-quality deterioration and improvement, respectively.

23

present as either bedded or domal salt; and (3) basinal waters have been forced through low-permeability shales (reverse osmosis), leaving a brine on the high-pressured side of the membrane. At present, there is no consensus about the relative importance of each of these three mechanisms to the origin of brine.

The theory of salt sieving (reverse osmosis) was first introduced by DeSitter (1947). During compaction, and in response to flow through tightly packed clays, anions are repelled from negative charges on clay-mineral surfaces (membrane effect), whereas positively charged ions may pass through the clay layers, maybe in a stepwise fashion moving from one cation exchange site to the next (Atkinson and others, 1986). This way, chloride as well as large, multivalent cations and heavy isotopes tend to be enriched on the high-pressure side of the membrane. Schwartz (1974) proposed that this process may also be active at relatively shallow depths in parts of Canada, involving residual effective stress conditions across till layers. According to Land (1987), opponents of the salt-sieving theory note that (a) sufficient pressure gradients may not be generated in nature, (b) the chemistry of brines in shale-rich sections doesn't appear to conform to expected patterns, and (c) a wide variety of formation-water types are observed in similar settings. Instead of the salt-sieving theory, Land (1987) suggested that the primary source of salinity are evaporite deposits. Evaporites have always been deposited through geologic time but have become increasingly rarer in progressively older rocks, suggesting dissolution. Garrels and MacKenzie (1971) estimated that present evaporites have been dissolved and precipitated 15 times in the last three billion years, resulting in an average cycle of 200 million years. Dissolution of salt in the subsurface may be the best explanation for (a) this relatively short cycle, (b) the missing salt in older rock formations, and (c) the abundance of sedimentary basin brines (Land, 1987). In addition to evaporite solution, mineral equilibria control the overall composition of brines. Sodium and calcium are by far the most dominant cations in almost all brines because equilibrium conditions between Na-feldspar (albite) and K-feldspar result in sodium dominance over potassium, and equilibrium conditions between calcite and dolomite result in calcium dominance over magnesium (Helgeson, 1972; Land, 1987). To explain other constituents in brines, Land (1987) proposed two types of rock-water interaction, one occurring in the sedimentary basin itself and the other occurring in the basement underlying the sedimentary basin.

The massive destruction of detrital feldspar releases significant amounts of calcium, potassium, strontium, barium, lithium, etc., to solution. Feldspar equilibria prove that large amounts of potassium cannot remain in solution until temperatures reach very high values, and potassium is commonly consumed in the formation of diagenetic illitic clay from a smectitic precursor (Schmidt and McDonald, 1979). The strontium and barium contained in brines are known to be vastly more abundant than predicted simply by the evaporation of seawater, and the $^{87}Sr/^{86}Sr$ ratio of the brine is commonly elevated. ^{87}Sr is produced by the decay of ^{87}Rb, an element characteristic of silicate phases, especially K-feldspar. Thus, significant involvement of silicate phases in determining brine chemistry is proven. Other isotope ratios, such as $^{18}O/^{16}O$ (Clayton and others, 1966) and D/H (Yeh, 1980)

24

are shifted from the values expected for surficial seawater-derived brines, providing additional evidence for extensive rock-water interaction in the subsurface.

It is well known that fluid inclusions in minerals of hydrothermal, metamorphic, or igneous origin are commonly very saline (Roedder, 1984). Metamorphism is accompanied by devolatilization, during which carbon dioxide and water are released, presumably to overlying sediments. The formation of slaty cleavage during low-grade metamorphism apparently requires the loss of large volumes of the rocks themselves (Buetner and Charles, 1982). The loss of appreciable volumes of insoluble components, such as SiO_2 and Al_2O_3 means that they must be transported into overlying strata. In addition to such large-scale material transport, the protons which are bound in alumino-silicates during weathering at the Earth's surface are progressively replaced by cations during metamorphism. For example, the H/Na of a solution in equilibrium with albite and kaolinite increases nearly two orders of magnitude over the temperature interval 25 to 200°C (Helgeson, 1972). In solutions dominated by chloride, weak HCl is thus produced. Acid water lost from metamorphic reactions into the overlying sediments will be neutralized by minerals like calcite, generating Ca-enriched solutions and CO_2. In fact, such a reaction sequence is one possible reason why the CO_2 content of natural gasses increases with increasing depth (Lundegard and Land, 1986). The $^{13}C/^{12}C$ ratio in natural gas also tends to become enriched in ^{13}C with increasing CO_2 content as more carbon is apparently derived from inorganic as opposed to organic sources.

Occurrences of brines in igneous rocks may be of multiple origin, as suggested by a long list of possible sources suggested in the literature. Edmunds and others (1987) list the following modes of possible origin: (1) marine transgression and subsequent concentration (for example, Frape and Fritz, 1982), (2) migration of sedimentary basin brines (for example, Fritz and Frape, 1982), (3) residual hydrothermal fluids (for example, Alderton and Sheppard, 1977), (4) dissolution of grain-boundary salts (for example, Grigsby and others, 1983), (5) silicate mineral hydrolysis and related water–rock interactions (for example, Edmunds and others, 1984, 1985; Frape and others, 1984), (6) breakdown of fluid inclusions in quartz and other minerals (for example, Nordstrom, 1983), and (7) radiolytic decomposition of water during α-series decay (Vovk, 1981). Some of these processes are believed to have caused brine concentrations of up to 550,000 mg/L TDS (Vovk, 1981). Edmunds and others (1987) added an extensive list of chemical reactions to the above list to explain the origin of saline ground water in a granite of Cornwall, United Kingdom. Eliminating ancient sea water as a potential source ($\delta^{18}O$ and δD indicated a local, meteoric origin), chemical reactions along the flow paths down to approximately 4,000 ft and at temperatures reaching 131°F (55°C) can explain these saline waters having TDS of up to 19,300 mg/L. Some results of these chemical reactions are: (a) acid hydrolysis of plagioclase and biotite as the prinicipal origin of ground-water salinity, (b) enriched Ca/Na ratios by selective reaction of the more calcic centers of zoned plagioclase, (c) high lithium concentrations related to biotite reaction, and (d) high Cl concentrations as a result of hydroxyl exchange for Cl in the biotite interlayer (Edmunds and others, 1987).

Evaporation from a shallow water table (within three or four feet of land surface) can lead to high salt concentrations in soils. This is known to have occurred in the San Joaquin Valley of California. There, ground-water salinity in Coast Range alluvium can be correlated with high concentrations of selenium, molybdenum, vanadium, and boron (Deverel and Gallanthine, 1989). Evaporation is the major process that accounts for high salinities in some closed basins in the western half of the United States. Weathering processes and selective mineral precipitations modify the chemical composition of closed-basin waters further. Major inorganic reactions in such settings are silicate hydrolysis, uptake of CO_2 from the atmosphere and/or of sulfate from oxidation of sulfides, and precipitation of alkaline earth compounds (Jones, 1965). Concentration ratios of major chemical constituents are much less uniform in closed-basin brines than in most deep-basin brines, the latter being nearly exclusively Na-Ca-Cl dominated. Closed-basin brines may be (1) chloride dominated, such as those in the Bonneville Basin, (2) carbonate dominated, such as those in Alkali Valley, Oregon, or (3) sulfate dominated, such as those of the Mojave Desert (Jones, 1965). These differences are due to differences in inflow characteristics and precipitation reactions although, in most instances, increased salinization (evaporation) is associated with a trend toward Cl dominance until halite precipitation is reached. Closed-basin brines start out by simply dissolving readily soluble mineral compounds, such as halite. Leaching of absorbed ions or of trapped interstitial fluids may be another process that provides dissolved minerals to the water. Silicate hydrolysis and subsequent evaporative concentration is the source of some of the high carbonate contents. Other processes that subsequently change the composition of the waters are: mixing with other water types, CO_2 addition by organic activity at lake bottoms or interstitially in lacustrine sediments, anaerobic decay of organic matter, and loss of sulfate (Jones, 1965). As the waters increase in concentration due to evaporation, precipitation reactions induce major changes to concentrations of individual ions. Calcium carbonate precipitates first, followed by gypsum, and finally by alkali salts. These processes can happen simultaneously, as described by Hunt (1960) and Jones (1965) for the Death Valley salt pan. There, carbonates precipitate in the outermost areas, sulfates in the intermediate zones, and chlorides at the center.

Ground-water evolution in closed basins may be similar to any other evolution along the flow path, from a low-TDS, Na-Ca-HCO$_3$ recharge water to a high TDS, Na-Cl water, as a result of reactions such as calcite dissolution and precipitation, cation exchange on clay minerals, and evaporation in discharge areas near the center of the basins. In discharge areas, mixing between this recharge water and brine may induce radical changes in the water type. Boyd and Kreitler (1986) demonstrated the geochemical evolution of a Ca-SO$_4$ recharge water to a Na-Cl brine in a gypsum playa in West Texas (Fig. 12). Brine evolution included precipitation of calcite, gypsum, and dolomite.

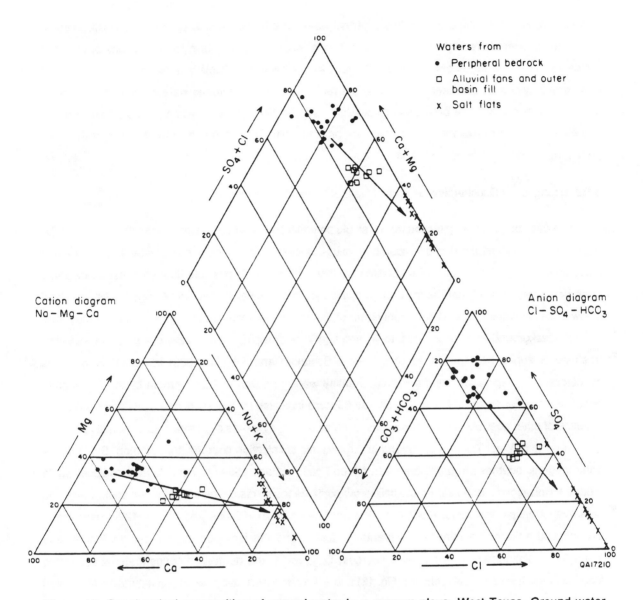

Figure 12. Geochemical composition of ground water in a gypsum playa, West Texas. Ground water evolves from a Ca-SO₄ type in recharge areas to a Na-Cl type in discharge areas (salt flat) (from Boyd and Kreitler, 1986).

In many areas of the United States, hydrothermal waters are found close to or at land surface. In some cases, these waters are high in mineral content as a result of increased mineral reactions at elevated temperatures and at large depths from which these waters originate. Stable isotope compositions in most geothermal systems have been found to reflect recharge of local meteoric waters, modified by one or more physical or chemical processes (Truesdell, 1976; in Welch and Preissler, 1986). One of these processes may be evaporation at land surface prior to infiltration, as reported by Welch and Preissler (1986) for Bradys Hot Springs, Nevada.

3.1.3. Examples of Geochemical Studies of Natural Saline Ground Water

In a case of salt-water intrusion from a single potential source of salt water, Sproul and others (1972) used chloride concentrations and temperature values to identify deep, artesian salt water as the source of intrusion in parts of Lee County, Florida. Upward flow in that area is possible through either abandoned, open boreholes or along natural fault and fracture systems. Similarly, Wait and Gregg (1973) identified inflow of saline water in the Glynn County area of Georgia by an increase in chloride concentration from average background levels of 23 mg/L to several hundreds of mg/L. There, heavy pumpage from shallow fresh-water zones was the cause of intrusion. Sowayan and Allayla (1989) identified progressive concentration (evaporation) as the source of saline water in a part of Saudi Arabia by plotting sodium concentrations versus chloride concentrations. This was indicated by a slope of near unity and the general absence of other possible salt-water sources, such as brines, evaporite deposits, or geothermal springs.

In parts of West Texas, natural and anthropogenic salinization sources are common. Some of the anthropogenic sources are associated with oil- and gas-field operations or agricultural techniques while natural salinization is associated with shallow evaporite deposits, discharge of saline formation water, and high evapotranspiration rates. Richter and Kreitler (1986a,b) and Richter and others (1990) used major cations and anions, the minor constituents Br and I, and the isotopes oxygen-18, deuterium, and sulfur-34 to distinguish between these sources. Oxygen-18 and deuterium concentrations separated local recharge from nonlocal recharge (Fig. 13) and Br/Cl and Na/Cl ratios separated natural halite solution from deep-basin discharge. These two sources also were differentiated clearly by other constituent ratios, such as I/Cl, Mg/Cl, K/Cl, Ca/Cl, and (Ca+Mg)/SO$_4$ (see also chapter 3.2). The same constituents have been used in Oklahoma and Kansas to differentiate natural halite solution from oil-field brines (for example, Ward, 1961, Leonard, 1964, Whittemore and Pollock, 1979, Gogel, 1981, and Whittemore, 1984, 1988). The concept of local and nonlocal recharge reflected in oxygen and hydrogen stable isotopes was also used by Banner and others (1989). Isotopic concentrations in saline waters of central Missouri are much lighter than typical for Missouri rain water, which suggested a source further to the west. According to Banner and others (1989), the Front Range of Colorado is the most likely area of recharge for these waters. This recharge water picks up high salt contents from Permian salts (indicated by Br/Cl ratios)

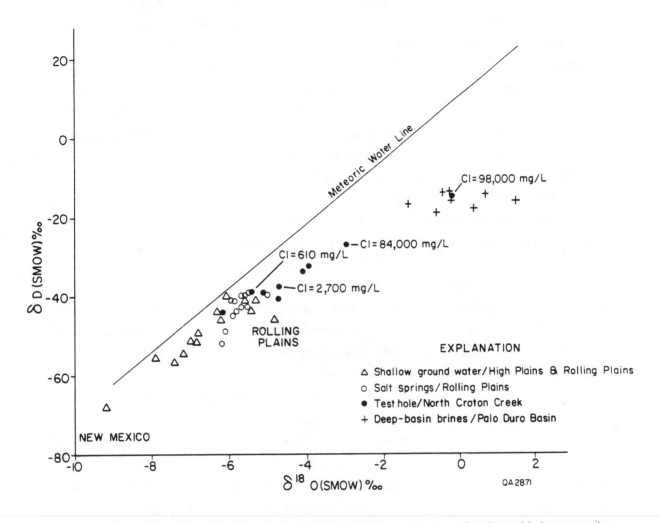

Figure 13. Separation of local recharge water from nonlocal ground water using the stable isotope ratios $\delta^{18}O$ and δD. Locally recharged ground water plots along the meteoric water line at the composition of local precipitation (open dots and open triangles), whereas nonlocally derived ground water typically is either enriched or depleted in isotopes compared to local precipitation (solid dots) (from Richter and Kreitler, 1986b).

underlying Kansas and finally mixes with fresh water near discharge areas in central Missouri. Ancient sea water or intensive interaction of ground water with aquifer material were precluded by these authors as possible sources of salinity because isotopic compositions of $^{87}Sr/^{86}Sr$ and ^{144}Nd (Neodymium) in ground water were significantly different from those in Paleozoic sea water and in local host rock. Bivariate plots of Br/Cl were also used by Morris and Bush (1986) in combination with plots of I/Cl and B/Cl to identify deep formation water as the source of salt-water contamination in parts of Arkansas (Fig. 14).

Lloyd and Heathcote (1985) demonstrated the use of strontium and iodine concentrations for differentiating saline ground water in the Lima Basin alluvial aquifer, Peru. Saline water derived from sea-water intrusion plotted on a dilution line between ocean water and alluvial water (Fig. 15). Waters that plot off this dilution line are associated with inflow from saline aquifers. One of these is characterized by Sr enrichment and the other is associated with iodine enrichment.

Saline playas and gypsum flats are natural discharge areas of closed basins in many parts of the western United States (Boyd and Kreitler, 1986). They also occur elsewhere, such as in the Murray Basin of southeastern Australia. Saline ground water in that area, ranging from 20,000 mg/L to 50,000 mg/L TDS, appears to be relict sea water with changes in concentrations reflecting local recharge/discharge conditions (Macumber, 1984). Concentrations increase due to evaporation at three major discharge points (Fig. 16a). Under these lake basins, ground water salinity also increases as the heavy brines recharge from the lakes. The similarity to sea water is preserved in these waters with little changes in concentration ratios. Brine under these lakes will continue to accumulate and spread beneath the less saline water in these basins until a massive inflow of fresh water changes the concentrations of the brine lakes (Macumber, 1984). Similarly, dissolved solids concentrations in saline springs and lakes on the Southern High Plains of Texas and New Mexico are the result of evaporation of shallow ground water rather than discharge of saline water from deep brine aquifers, as indicated by constituent ratios of Cl/Br, Na/K, and Cl/SO$_4$ being consistently smaller in evaporated waters than in deep-basin brines (Wood and Jones, 1990). Salinities increase in lakes and springs as surface water evaporates, and percolation water is recycled from the lakes to the springs (Fig. 16b). Locally, ground water in the Ogallala Formation of the Southern High Plains is affected by these evaporated waters, but also by oil-field brines. Salinization by these two sources can be differentiated through the use of salinity diagrams (Fig. 17), as evaporative waters are characterized by Na-SO$_4$ waters in contrast to high-TDS, Na-Cl facies typical for oil-field-brine affected water (Nativ, 1988).

The distribution of natural salinity in shallow ground water reflects the geomorphology and hydrology of the system, such as in alluvial fans in the western San Joaquin Valley of California (Deverel and Gallanthine, 1989). Lowest salinities were found in the upper and middle fan areas and highest salinities (greater 5,000 mg/L TDS) in the lower fan areas and at the margins of the fans where fine-grained soils are poorly drained. Historically, soils associated with ephemeral streams received less water and less flushing than larger intermittent streams and, therefore, were more saline. Leaching of soil salinity by irrigation

30

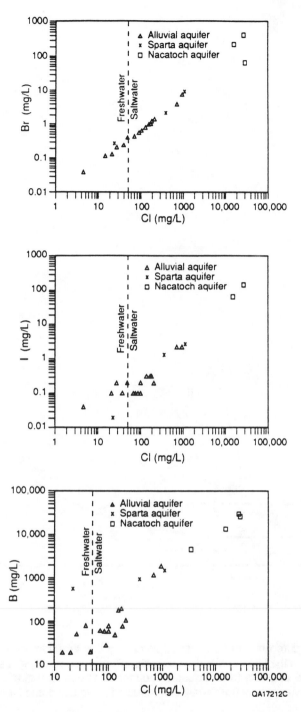

Figure 14. Use of bromide, iodide, boron, and chloride concentrations in bivariate plots for identification of salt-water sources. Mixing between fresh water and salt water is suggested by linear trends between potential endmembers of mixing (from Morris and Bush, 1986).

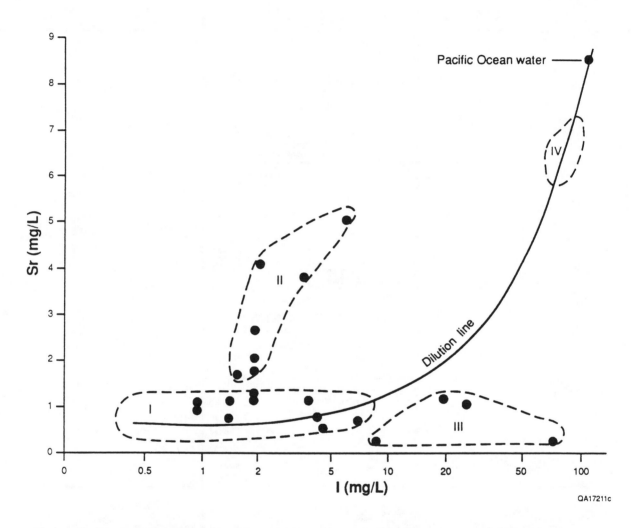

Figure 15. Separation of different ground-water origins using the strontium-iodide relationship in ground water. Data shown are from the Lima Basin alluvial aquifer, Peru: Group I water is associated with alluvium; Group II water enters the alluvium from Jurassic sediments; Group III water enters the alluvium from granodiorites; Group IV water results from sea-water intrusion (from Lloyd and Heathcote, 1985).

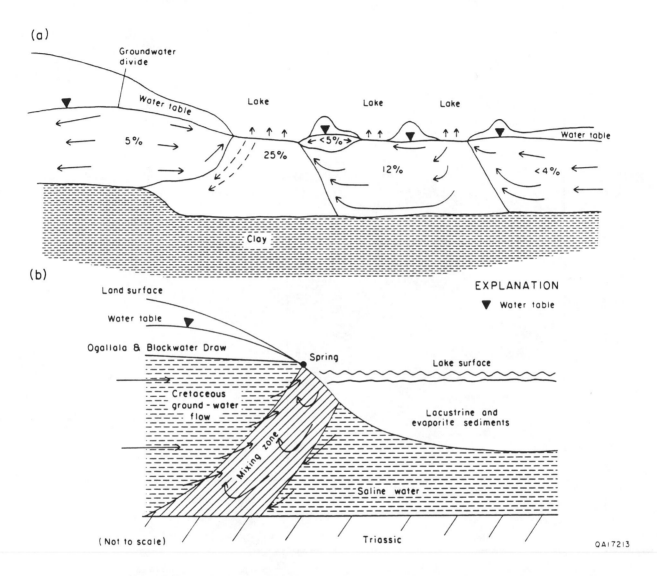

(a)

Groundwater divide

Water table

Lake   Lake   Lake

5%

Water table

25%

<5%

12%

<4%

Clay

(b)

Land surface

Water table

EXPLANATION

▼ Water table

Ogallala & Blackwater Draw

Spring

Lake surface

Cretaceous ground-water flow

Lacustrine and evaporite sediments

Mixing zone

Saline water

(Not to scale)

Triassic

QAI7213

Figure 16. Ground-water flow directions (arrows) and salinities (in percent) in parts of (a) the Murray Basin, southwest Australia, (from Macumber, 1984) and (b) at saline lakes, Southern High Plains of Texas (from Wood and Jones, 1990). Evaporation at major discharge sites and mixing causes TDS increases along the flow path.

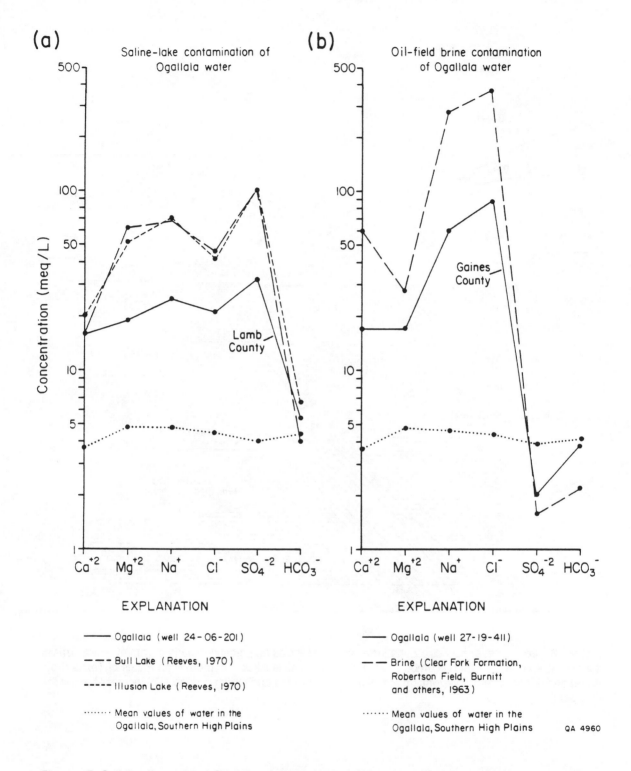

Figure 17. Salinity diagrams of Ogallala water contaminated by (a) saline lake water and (b) by oil-field brines (from Nativ, 1988).

water was the primary process that initially formed shallow, saline ground water. Subsequently, evaporative concentration caused further increases in salt content. This evaporative trend is illustrated in a plot of $\delta^{18}O$ and δD, which shows that samples collected at the lower fan areas are the most enriched ones (Fig. 18a). Waters with the highest salinities in the area are of the $Na-SO_4$ type (Fig. 18c), with sulfate concentrations controlled by gypsum precipitation, as suggested by saturation indices (Fig. 18b).

Geothermal spring water in Churchill County, Nevada, is of local, meteoric origin as indicated by $\delta^{18}O$ and δD values (Fig. 19) (Welch and Preissler, 1986). Salinization caused by evaporation is suggested by a shift toward oxygen and hydrogen isotope enrichment. This evaporation trend was also indicated by a uniform Br/Cl ratio. Of the minor constituents Ba, B, Br, F, Pb, Li, Mn, and Sr tested, bromide is largely controlled by evaporative concentration, whereas barium, fluoride, lead, and manganese are controlled by mineral phases such as barite, fluorite, cerussite, and rhodochrosite, respectively. Boron, lithium, and strontium may be indicators of the presence of hydrothermal waters, as concentrations are higher in the higher temperature samples (Welch and Preissler, 1986).

Saline ground water trapped in portions of aquifers since deposition or since subsequent salt-water intrusion generally does not contain any tritium because of the tritium isotope's short half-life of only 12.4 years. Anthropogenic sources of salt, such as road salt, in contrast, may contain measurable amounts of tritium, as these salts are dissolved and flushed into aquifers by modern precipitation. This relationship was used by Snow and others (1990) to distinguish trapped sea water from road-salt contamination in the coastal wells of Maine.

3.1.4. Significant Parameters

Most of the salinity sources described in this report occur naturally at some place or another, where they mix with fresh ground water. In other cases, mixing of naturally saline water with fresh water is initiated or facilitated by anthropogenic activities, such as heavy pumpage of fresh water, drilling through fresh-water and salt-water bearing zones, or disposal of produced water. In most instances, chemical characteristics will not differ significantly between natural mixing of fresh and salt water and artificial mixing of the same salt water with fresh water. Therefore, significant parameters for identification of natural salinization are the same as those for any individual source discussed in the following sections.

Salinization is generally indicated by an increase in chloride concentration. If this increase is substantial and occurs suddenly and is localized, a nonnatural mechanism and source should be suspected. However, if the change is subtle and of regional scale, a natural mechanism or source may be responsible.

The stable isotopes Oxygen-18 and deuterium are generally useful to distinguish between local precipitation water and water that is derived from a nonlocal source and identify evaporation of local recharge water. Molar ratios of major chemical constituents, such as Na/Cl, Ca/Cl, and Mg/Cl, can be used

35

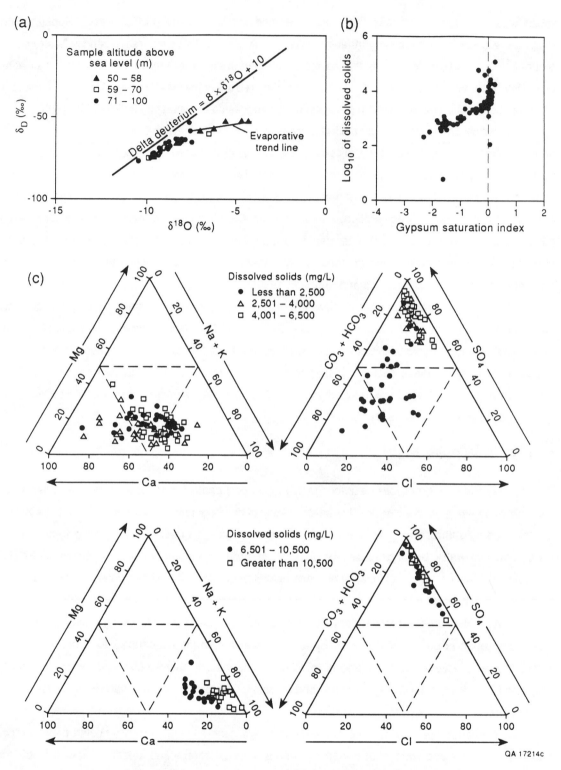

Figure 18. Natural salinization of ground water in parts of the San Joaquin Valley of California, controlled by (a) evaporation and (b) gypsum saturation, resulting in (c) a shift from low-TDS mixed-cation mixed-anion-type water to high-TDS, Na-SO$_4$ water (from Deverel and Gallanthine, 1989).

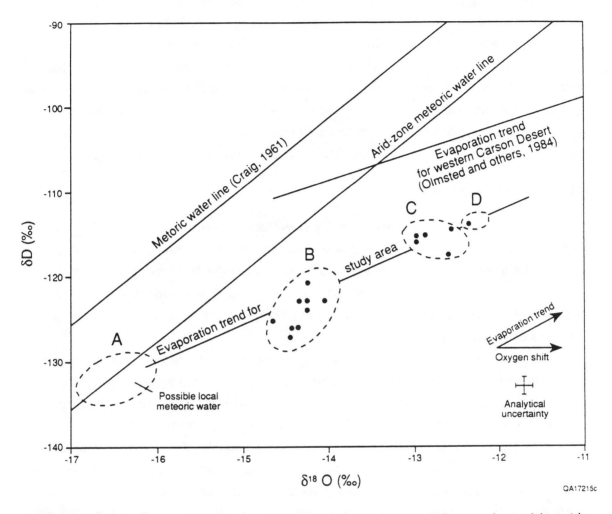

Figure 19. Stable-isotope composition of ground water affected by evaporation, causing enrichment in oxygen-18 and deuterium as salinity increases. Chloride ranges: A <50 mg/L; 910 <B <1,600 mg/L; 3,500 <C <3,700 mg/L; D = 4,200 mg/L (modified from Welch and Preissler, 1986).

to differentiate an evaporation trend (1:1 slope) from a mixing trend (typically not a 1:1 slope). Mixing trends can be evaluated best using the most conservative constituents dissolved in ground water, that is, chloride and bromide. These constituents are often useful to not only identify the mixing source of salinity (see also chapters 3.2 through 3.7), but also to estimate the mixing ratio (see also chapter 6.1).

Mixing of fresh water with naturally saline ground water or evolution of ground water toward higher salinities, as opposed to mixing with road-salt solutions, mixing with brine along boreholes, or disposal of produced oil-field brine, can be expected to be a relatively slow process during which the water has time to react extensively with the aquifer matrix. Therefore, saline ground water in its natural environment will reflect conditions of chemical equilibrium more closely than artificially induced mixtures of fresh water and saline water. This may be used to distinguish natural mixing or evolution from induced mixing.

3.1.5. State-by-State Summary

Natural saline ground water underlies nearly every state at some variable depths (Fig. 1). Where fresh-water aquifers are plentiful or where ground-water usage is low, little may be known about the base of fresh water. In other areas, in contrast, lack of usable water and, with it, heavy drilling in an attempt to find water has helped to define the interface between fresh and saline water remarkably well. The following sections taken entirely from miscellaneous published sources deal with the occurrence of saline ground water on a state-by-state basis.

Alabama: Saline ground water is found at shallow depths in the north, north-central, and southwesternmost parts of the state. TDS exceeding 3,000 ppm occur in southern Lowndes County, southwest of Montgomery, and in western Clarke and northeastern Washington Counties, including a well reported to contain between 10,000 and 35,000 ppm in southeastern Choctaw County (Feth and others, 1965). Depth to saline ground water exceeds 1,000 ft in the remainder of the state, including the Gulf Coast (Feth and others, 1965). Natural salinization in southwest Alabama is due to brine migration along faults, affecting all major aquifers in that area (Slack and Planert, 1988). In Green County, wells in the Gordo and McShan Formations furnish water ranging from 4 to 3,700 mg/L Cl and 3.8 to 2,560 mg/L, respectively. Many of these wells are flowing at land surface.

Arizona: Shallow saline ground water (TDS>1,000 ppm) is found in the northeastern and southwestern part of the state, including TDS greater than 35,000 ppm at depth of 500–1,000 ft in the Puerco River drainage east of Holbrook. TDS in ground water between 1,000 and 10,000 ppm occur along the Gila River, and between 1,000 and 3,000 ppm along the Colorado River of southwestern Arizona at depths less than 500 ft below land surface (Feth and others, 1965).

Water quality in many streams deteriorates in downstream direction. Often, mineral springs contribute large amounts of dissolved solids to the streams, such as those near the mouths of Chevelon and Clear Creeks near Winslow, where TDS concentrations of up to 4,000 ppm are recorded in the Little

Colorado River. Concentration greater than 6,000 ppm have been measured in the Gila River, in part due to inflow of saline tributary water, such as those from San Francisco River, which in turn receives saline water of up to 9,000 ppm from Clifton Hot Springs (Krieger and others, 1957). Over 200 tons of dissolved solids are discharged daily into the Salt River by three major springs that originate in the Mogollon Rim in central and east-central Arizona (Fuhriman and Barton, 1971). At Phoenix, the Salt River carries approximately 500,000 tons of salt per year (Skibitzke and others, 1961).

Most of the ground-water basins in the Salt River Valley are overdeveloped and yield water which is mineralized, with TDS exceeding 3,000 mg/L in many wells. Highest salt concentrations are often found in the center of heaviest ground-water usage; maximum values exceed 7,000 mg/L. The deeper aquifers usually are of better water quality than the shallow aquifers. Improperly abandoned deep wells can lead to pollution of these deep fresh waters by shallow saline water depending on the head difference between them. Mineralization is high in the San Pedro, Willcox, and Safford subbasins within the Upper Gila River and adjoining Mexican drainage. TDS as high as 5,000 ppm have been reported from wells in the Palomas Plains. Highly mineralized ground water in the Colorado River Basin of southwestern Arizona can be found in the Ranegrass Plain (TDS up to 4,000 ppm) and in some wells in the Yuma area (TDS up to 5,000 ppm) (Fuhriman and Barton, 1971). In the Safford Valley, ground-water salinity has increased from 1940 through 1972 as a direct result of predominantly pumping-induced aquifer leakage; other factors contributing to the salinity are natural recharge to the water table by saline water, vertical flow of artesian ground water through saline lacustrine beds, and agricultural recharge waters (Muller, 1974).

Natural mineralization in areas of low precipitation and naturally restricted drainage is the most common ground-water quality problem in Arizona. Leaching of highly saline soils and rocks accounts for most of the salinity problems (Smith, 1989). Of the total irrigable area of 1,565,000 acres, approximately 25 percent (398,830 acres) were considered saline or alkaline in 1960 (Fuhriman and Barton, 1971).

Arkansas: Saline ground water (TDS>1,000 ppm) underlies Arkansas at shallow depth (<500 ft below land surface) within a narrow zone that extends from the northeastern to the southwestern part of the state. To the east of this zone, saline ground water is restricted to depths mostly greater than 1,000 ft below land surface. Another shallow zone of saline water is located in westernmost Arkansas including all of Sebastian and Crawford Counties and parts of Logan and Franklin Counties (Feth and others, 1965).

Lateral movement of saline water into fresh water by updip migration resulting from pumping was detected in areas of eastern and in the Sparta Aquifer in southern Arkansas (Newport, 1977; U.S. Geological Survey, 1984). In the Brinkley area, approximately 56 mi^2 of the alluvial aquifer have been contaminated by salt water from underlying saline formations (Morris and Bush, 1986). Artesian upward movement of salt water in this area is enhanced by irrigation and water-supply pumping. A chloride concentration greater than 50 mg/L was considered indicative of salt-water contamination and was used for mapping of contaminated areas by Morris and Bush (1986).

39

California: Isolated areas of shallow saline ground water (TDS>1,000 ppm at depth less than 500 ft below land surface) occur in many parts of the state, including major areas such as the San Joaquin Valley, the Colorado River Basin, and most of Imperial County. TDS in excess of 10,000 ppm are reported from closed basin lakes in the southeast (Feth and others, 1965). Saline lakes in California include from north to south (located mostly along the California-Nevada/Arizona state lines): Surprise Valley, Honey Lake, Pyramid Lake, Lake Tahoe, Mono Lake, Saline Valley, Owens Lake, Searles Lake, Soda Lake, Bristol Dry Lake, Cadiz Lake, Danby Lake, and Salton Sea (Hardie and Eugster, 1970).

Approximately 33 percent (3,745,000 acres) of California's total irrigable land was considered saline or alkaline in 1960. Maximum TDS concentrations in mg/L at selected locations throughout California as reported through 1965 were: Ukiah Valley, 1,280; Santa Rosa, 560; East Bay (San Francisco), 4,100; South Bay (San Francisco), 1,750; Livermore Valley, 4,700; Petaluma Valley, 19,760 (sea-water intrusion); Napa Valley, 1,840; Sonoma Valley, 660; Suisum-Fairfield Valley, 2,560; Pajaro Valley, 1,310 (TDS>5,000 in areas of sea-water intrusion); Gilroy-Hollister Basin, 1,480; San Luis Obispo Hydrologic unit, 3,024; Carrizo hydrologic unit, 10,460; Santa Maria-Cuyama hydrologic unit, 5,088; San Antonio hydrologic unit, 4,070; Salinas Valley, 3,134; Carmel Valley, 729; Paso Robles Basin, 3,280; Santa Ynez hydrologic unit, 21,800; Santa Barbara hydrologic unit, 2,487; Mission Basin, 13,930; San Dieguito Valley, 27,402; Tia Juana Valley, 4,680; Oxnard Plain area, 33,180; West Coast Basin, 41,397; East Coastal Plain, 41,800; Main San Gabriel Valley, 1,140; Chino Basin, 1,417; Perris Valley, 11,620; San Joaquin County, 3,840; San Joaquin Basin, 6,400 (connate water?); and Tulare Basin, 6,450 (Fuhriman and Barton, 1971).

Mineralized hot springs contribute to the salinity in the North Lahontan Basin in the Bridgeport area, whereas intrusion of brackish water in gravel-packed wells open to zones of variable water quality may pose a problem in some wells in the San Joaquin Valley (Fuhriman and Barton, 1971). Dissolved solids in shallow ground water from alluvial deposits along the San Joaquin River in Fresno County vary between less than 1,000 mg/L to greater than 15,000 mg/L. Natural salinization of soils occurred by evaporation of ground water at times when the water table was close to land surface. These salts were subsequently dissolved by irrigation waters, causing highest concentrations of TDS in those areas that have been irrigated only in the last 40 years (Deverel and Gallanthine, 1988).

To compensate for extensive overdraft of ground water in Orange County during the 1940's, large quantities of Colorado River water were imported and recharged along the Santa Ana River channel. This resulted in salinization of a substantial portion of ground water in the county due to the high TDS content of the imported water (U.S. Environmental Protection Agency, 1973).

Saline water intrusion in southern Alameda County is due to the following mechanism: direct movement of bay waters through natural "windows" (that is, high-transmissive zone in an otherwise low-transmissive clay layer), spilling of degraded ground waters through windows in clay layer, slow percolation of salt water through reservoir roof (clay layer), and spilling or cascading of saline waters or degraded ground water through deep wells (U.S. Environmental Protection Agency, 1973).

Colorado: Saline ground water (TDS>1,000 ppm) occurs at shallow depth (<500 ft below land surface) in the western third and in southeastern Colorado. Ground waters containing TDS greater than 3,000 ppm underlie the state at depths of >1,000 ft in the eastern and northeastern parts (Feth and others, 1965).

Approximately 35 percent (982,000 acres) of Colorado's irrigable land was considered saline or alkaline in 1960 (Fuhriman and Barton, 1971).

Approximately 37 percent of the salt load in the upper part of the Lower Colorado River Basin is contributed by irrigation-return flow. The other 63 percent is contributed by interaction of water with naturally saline soils and aquifer material overlying a marine shale that contains numerous lenses of salt. Within an area comprising 1,876 ha, approximately 51,000 metric tons of salt are added to the river system by these two sources (Skogerboe and Walker, 1973).

Delaware: Saline ground water (TDS>1,000 ppm) underlies Delaware at depth >1,000 ft except for the coastal strip in the southeast, where saline ground water may be found at depth <500 ft below land surface (Feth and others, 1965). The inland limit of saline ground water is located along the coastline in Cretaceous rocks, falls along the western and northern borders of Sussex County in Tertiary rocks, and stretches all the way north to the city of Dover in Pleistocene formations (Miller and others, 1974).

Brackish water occurs locally in shallow water-table aquifers in coastal areas. Some salt water is contained in the Chesapeake Group (Miocene) in Sussex County, and the interface between fresh and saline water in the Piney Point Formation (Eocene) is located just north of Milford, extending to the northeast across the state. Brackish water also exists in the Magothy and Potomac Formations (Cretaceous) a few miles south of Middletown. Heavy pumpage near these brackish-water zones may cause water-quality deterioration due to inflow of saline water toward the wells. High chloride concentrations between 6,000 mg/L and 17,000 mg/L were measured in core samples obtained from wells along the Atlantic Coast of southeastern Delaware. The depths of wells from which core samples were obtained ranged from 5 ft to 60 ft below land surface (Woodruff, 1969).

Florida: A 30-mile-wide stretch of shallow saline ground water (TDS>1,000 ppm at depth <500 ft below land surface) extends from St. Augustine in the north to Vero Beach in the south along the Atlantic Coast of Florida. An isolated occurrence of shallow saline ground water is also reported for westernmost Florida. In contrast, saline ground water generally is found at depths between 500 and 1,000 ft in southern Florida and deeper than 1,000 ft in the remainder of the state. In southern Florida, large quantities of moderately saline water with TDS less than 5,000 mg/L can be found in the Avon Park Limestone Formation at depths greater than 1,200 ft below land surface. Less saline water at shallower depths, between 300 and 1,100 ft below land surface, occurs in the Hawthorn, Tampa, and Suwannee Formations (Meyer, 1971). Individual wells along the western Gulf Coast in Sarasota and Manatee

Counties as well as wells along the southeastern shoreline may exhibit salinities in excess of 3,000 mg/L TDS (Feth and others, 1965).

The Florida State Geological Survey collected information on 1,800 uncontrolled flowing wells between 1955 and 1959. Chloride concentration in approximately 50 percent of the wells exceeded the drinking water limit of 250 mg/L. At present, there are an estimated 2,000 to 3,000 wells that flow poor-quality water in southwest Florida alone. At La Belle, upward leakage through five deep artesian wells that were not in use and lateral movement through a subsurface water-distribution system were the probable sources of contamination of the shallow aquifer over a broad area (Miller and others, 1977).

The chemical composition of ground water in the Floridan aquifer of coastal southwest Florida changes from a fresh Ca-Mg-HCO_3 type water in the upper, upgradient areas to a similar fresh Ca-Mg-SO_4 type water in the deeper, more permeable, dolomitic zone, and finally to a highly saline Na-Mg-Cl type water in the downgradient areas. Water-quality degradation can occur in areas of high pumpage, such as the Verna well field in northern Sarasota County (Steinkampf, 1982).

Vertical migration of saline water along a deep, abandoned irrigation well at Highland Estates caused chloride increases in the shallow aquifer and in some water wells from background levels of 20 mg/L to contamination levels of 590 mg/L (Boggess, 1973; Miller and others, 1977).

Georgia: Saline ground water (TDS>1,000 ppm) underlies the southern half of the state at depths exceeding 1,000 ft (Feth and others, 1965).

Locally, salinization occurs due to flowing, abandoned wells (Clarke and McConnell, 1988). Intrusion of saline water at Brunswick is caused by upward migration of brines through faults and fractures in response to the daily withdrawal of 105 million gallons by industry and municipalities. In one test well, chloride concentration increased from 4,400 to 7,000 mg/L during the past 10 years (U.S. Geological Survey, 1984).

Idaho: Isolated occurrences of shallow saline ground water (TDS >1,000 ppm at depth <500 ft) are reported from wells in the Snake River Basin of southern Idaho and in the southern part of Oneida County in southeast Idaho (Feth and others, 1965).

Thermal water occurs at many locations at depths greater than 400 ft below land surface. Maximum chloride concentration in nonthermal valley-fill aquifer is 3,900 mg/L (Parliman, 1988).

Illinois: Saline ground water of TDS>1,000 ppm underlies central and southern Illinois at depths less than 500 ft below land surface (Feth and others, 1965). Saline water from the Mt. Simon aquifer may leak into shallow fresh-water aquifers due to overpumping (Atkinson and others, 1986).

Indiana: The entire state of Indiana is underlain by saline ground water (TDS >1,000 ppm) at depths less than 500 ft below land surface. Individual wells have been reported supplying water containing between 3,000 and 10,000 ppm of TDS (Feth and others, 1965).

Ample rainfall in the state prevents overpumping of shallow aquifers which reduces the potential of salt-water intrusion. However, a leaky aquifer in Knox County has caused salinization of fresh ground water

(Atkinson and others, 1986). In the Mt. Vernon–West Franklin area, upward flow of saline water through fault zones into fresh water was caused by overpumping of the fresh-water aquifer (Newport, 1977). Near Vincennes, saline water has moved up through bedrock fractures into the bottom of the glacial outwash aquifer near the municipal well field, endangering future water supplies if pumpage increases more than fourfold (Shedlock, 1978).

Iowa: Except for the northeastern part, all of Iowa is underlain by saline (TDS between 1,000 to 3,000 ppm) to very saline (TDS between 10,000 to 35,000 ppm) ground water at depths less than 500 ft below land surface (Feth and others, 1965). Despite this shallow occurrence of saline ground water, Iowa does not currently experience salt-water intrusion, but the potential of contamination from shallow saline water in response to intensive pumping exists (Newport, 1977; Atkinson and others, 1986). In glacial drift aquifers underlying Wayne County, TDS concentrations range up to 3,600 mg/L in wells as little as 100 ft deep (Cagle, 1969).

Kansas: The eastern half of the state is underlain by ground water containing TDS greater than 1,000 ppm at depths less than 500 ft below land surface. Concentrations in excess of 10,000 ppm occur locally, and are widespread in central and south-central Kansas (Feth and others, 1965). Natural salinization of Kansas' surface water is derived principally from salt springs, salt marshes, and direct contact with saliferous geologic formations. Major sources of saline water occur in a 150-mile-wide band that stretches across the central part of the state from the southern to the northern border (Krieger and others, 1957). Salt-water intrusion has been reported in Sedgwick, Reno, Harvey, McPherson, Saline, Dickinson, Seward, and Meade Counties. Salinization problems are especially serious in outcrop areas of the Wellington Formation along the Solomon and Smoky Hill Rivers (Atkinson and others, 1986).

Natural intrusion of mineralized water is occurring in central Kansas from Permian Redbeds, the Wellington Formation, and the Dakota and/or Kiowa Formations. A prominent example of chloride and sulfate intrusion from the Dakota and or Kiowa Formations is along the Smoky Hill and Saline Rivers in Russell County. The Wellington Formation contains abundant beds of gypsum and a thick bed of halite which pose problems from Ottawa-Dickinson Counties to the south. The salt bed (Hutchinson Salt Member, up to 350 ft thick) is dissolved easily by circulating ground water. The most damaging effects from salt dissolution are along the Smoky Hill and Solomon Rivers in eastern Saline and western Dickinson Counties and the Belle Plain area in Sumner County on the Ninnescah and Arkansas Rivers and the Slate and Salt Creeks. Certain zones of the Permian Red Beds contain salt and gypsum which is easily dissolved by circulating ground water. Surface streams and ground water contaminated by this source are found along the lower reaches of the Rattlesnake Creek in Stafford and Reno Counties and along the South Fork Ninnescah River in eastern Pratt and western Kingman Counties. Ten minor areas of salt-water intrusion occur along (1) the Solomon River in Mitchell and Cloud Counties (for example, Waconda Spring west of the city of Beloit), (2) Salt Creek in Mitchell and Lincoln Counties, a tributary of the Solomon River, (3) Buffalo Creek in Jewell and Cloud Counties, (4) Salt Creek and Marsh Creek in Republican County,

(5) the Cottonwood River in Marion County, (6) the Medicine Lodge River in Barber County, (7) the Salt Fork Arkansas River in Comanche and Barber Counties, (8) the Cimarron River in Meade County, (9) the Crooked Creek in Meade County, and (10) Cow Creek in Reno and Rice Counties (Hargadine and others, 1979).

In the area south and west of Salina, fresh water moves down through collapsed and brecciated shales to recharge the Wellington aquifer, dissolving halite within the Hutchinson Salt Member. The resulting brine then moves northward beneath the Smoky Hill River Valley to the Saline River Valley. Upward discharge of brine occurs through collapse structures in an area from about three miles east of Salina to just east of Solomon, where the potentiometric surface of the Wellington aquifer is above the water table in the alluvial aquifer. According to landowners, the interface between fresh water and saline water in the alluvial aquifer along the Smoky Hill River in northern Saline County is about 35 ft below land surface. Chloride concentrations of 48,000 mg/L at 94 ft depth and of 180,000 mg/L in gypsum cavities at 112 ft have been reported. A relatively impervious shale separates the shallow fresh-water unit from the deeper brine unit except for areas of collapse and brecciation caused by salt and gypsum dissolution (Gillespie and Hargadine, 1981). Major rivers affected by these saline ground-water discharges include the Smoky Hill near Salina, the Arkansas, Ninnescah, Saline, and Solomon Rivers. Increased pumpage from the principal alluvial aquifer in central Kansas ("Equus beds") could cause inflow of mineralized water from the Arkansas River into the aquifer (U.S. Geological Survey, 1984).

Approximately 24 percent (102,000 acres) of all the irrigable land in Kansas was considered saline or alkaline in 1960 (Fuhriman and Barton, 1971).

Kentucky: Most of Kentucky, with the exception of its easternmost part, is underlain by shallow saline water, that is, ground water containing TDS concentrations greater than 1,000 ppm underlie the state at depths less than 500 ft below land surface (Feth and others, 1965).

Louisiana: Shallow saline ground water (TDS>1,000 ppm) can be found along the Gulf Coast and in northeast and central Louisiana (Feth and others, 1965). Sea-water intrusion has occurred all along the coastal shores of Louisiana. In addition, the cities of Lake Charles, Baton Rouge, and New Orleans have experienced severe cases of salt-water intrusion due to high pumpage of ground water. Shallow saline water appears to be associated with salt-dome provinces. Shallow saline water also occurs along the northeastern part of the state in the Mississippi River alluvial aquifer (Atkinson and others, 1986).

During periods of low flow tidal water invaded fresh-water aquifers in the Vermillion River area (Newport, 1977). In southwestern Louisiana, saline water is moving landward and updip in the Chicot aquifer, and, in the Baton Rouge area, saline water is moving across faults toward municipal wells (Whiteman, 1979; U.S. Geological Survey, 1984). Shallow salt water also occurs in Pleistocene delta sediments that may not yet be flushed.

Maine: Overpumping has resulted in intrusion of tidal estuary waters into local aquifers south of Augusta, in Kennebec and Sagadahoc Counties. Several domestic water wells have been affected by salt-water intrusion whereby the problem fluctuates with time over the year (Atkinson and others, 1986).

A 300-ft-deep well producing from the bedrock aquifer near the town of Bowdoinham, Sagadahoc County, was contaminated by salt water from the tidal reach of the Kennebec River, resulting in abandonment of the well (Miller and others, 1974).

Maryland: Shallow saline ground water (TDS>1,000 ppm) occurs in the southern half of the Chesapeake Bay peninsula (Feth and others, 1965). The inland limit of saline ground water in coastal plain aquifers are along the coast for the Cretaceous aquifer, along a line from Dorchester to the southwestern state-line corner with Delaware in the Tertiary aquifer, and along a line parallel to the Chesapeake shore along the shore in the south and approximately half way between the shore and the Maryland-Delaware state line in the north, crossing the state line just north of 39 degrees latitude (Miller and others, 1974). Downdip portions of coastal aquifers contain natural salt water (Wheeler and Maclin, 1988).

Michigan: A zone of moderately saline ground water (TDS between 3,000 and 10,000 ppm) underlies most of the eastern part of the state along Lake Huron to Lake Erie. Saline ground water (TDS>1,000 ppm) underlies the remainder of the Michigan peninsula at variable depths of less than 500 ft to more than 1,000 ft below land surface. Saline ground water is also encountered at shallow depth along Lake Michigan and the western shore of Lake Superior (Feth and others, 1965). Higher TDS in ground water occurs in the eastern part of the state, where more water is produced from bedrock aquifers. The average depth to saline ground water along Lake Erie, Lake St. Clair, Lake Huron, and near Saginaw Bay is 200 ft.

Pumpage has resulted in upward intrusion of saline water from deep bedrock into glacial aquifers at various places throughout the state. This problem has been aggravated locally by leaky well casings (Newport, 1977). Saline water in predominantly shale and silty shale of the Saginaw Formation underlies glacial deposits at shallow depths in the area of Bay County, in the east-central part of Michigan's Lower Peninsula. Chloride concentrations increase markedly to several thousand mg/L in many wells deeper than 100 ft below land surface. Abandoned coal mines in the area may contribute to mixing of deep, saline water and shallow, fresh water, but do not constitute a major source of water-quality deterioration (Twenter and Cummings, 1985).

Many salt seeps and salt springs occur in the state (Krieger and others, 1957). Natural brines are produced from rocks of the Detroit River Group at Manistee and Ludington in the western part of the state, and at Midland, St. Louis, and Mayville in the eastern part of the state (Sorensen and Segall, 1973).

Minnesota: Shallow saline ground water (TDS>1,000 ppm) has been reported along the western border, from the northeastern part (along the shore of Lake Superior), and the southwestern and southeastern parts of the state (Feth and others, 1965; Albin and Breummer, 1988). In northwestern Minnesota, mineralized ground water discharges from Ordovician and Cretaceous bedrock. Pumping from glacial aquifers in this area has caused upward intrusion of saline water from these deep bedrock aquifers (Newport, 1977). Natural discharge occurs from the Red River-Winnipeg aquifer, degrading water quality in overlying alluvial deposits and in the Red River of the North. Chloride concentrations of up to 46,000 mg/L have been reported from wells along the shore of Lake Superior (Albin and Breummer, 1988).

Mississippi: Saline ground-water generally underlies the state at depths of more than 1,000 ft below land surface (Feth and others, 1965). Geohydrologic data indicate that most of the principal aquifers of the state previously contained salt water. This salt water was later partially replaced by fresh water but still occupies downdip portions of these aquifers (Bednar, 1988).

Missouri: Saline ground water containing more than 1,000 ppm TDS underlies the northern one-third of the state at depths less than 500 ft below land surface. This includes an area of very saline water (TDS from 10,000 to 35,000 ppm) that stretches from Clay County in the west to Marion County in the east (Feth and others, 1965). The intrusion of natural salt-water from saline formations into shallow fresh water is spreading from the northwestern part of the state toward the south (Atkinson and others, 1986). Intrusion of brackish water in response to ground-water withdrawal has occurred in Bates, Barton, and Vernon Counties (Carpenter and Darr, 1978).

The history and flow path of saline ground water in central Missouri has been suggested to be (1) meteoric recharge in the Front Range of Colorado, (2) dissolution of Permian halite in the subsurface of Kansas, (3) interaction with predominantly silicate mineral assemblages in Paleozoic strata, (4) dilution and migration to shallow aquifer levels in central Missouri, and (5) mixing with local meteoric recharge through Mississippian carbonates (Banner and others, 1989).

Montana: Most of the eastern half of the state is underlain by saline water (TDS>1,000 ppm) at depths less than 500 ft below land surface. Locally, TDS concentrations exceed 10,000 ppm (Feth and others, 1965). Maximum TDS concentrations in the glacial-deposits aquifer are 30,000 mg/L and in the Virgelle aquifer 5,100 mg/L (Taylor, 1983).

Nebraska: Saline ground water (TDS>1,000 ppm) underlies the state at 500 to 1,000 ft below land surface in the southeast and at greater than 1,000 ft below land surface in the remainder of the state. TDS increase to values greater than 3,000 ppm in the southwest (Feth and others, 1965).

Salt water appears to be no serious threat to ground water in the state except for the easternmost part where TDS up to 3,500 mg/L are reported from the Dakota Aquifer and Paleozoic rocks (Engberg and Druliner, 1988). Taylor (1983) reported maximum TDS concentrations in the Dakota aquifer of 30,000 mg/L.

According to Atkinson and others (1986), local occurrences of salt-water intrusion have been reported in Saunders, Lancaster, and Saline Counties. Saline surface water occurs in some of the sandhills lakes in western Nebraska and in some localities in eastern Nebraska (Krieger and others, 1957).

Approximately 24 percent (290,000 acres) of Nebraska's total irrigable acreage were considered saline and alkaline in 1960 (Fuhriman and Barton, 1971).

Nevada: Saline surface water and shallow saline ground water has been reported locally throughout central and western Nevada. TDS concentrations vary between 1,000 ppm and 35,000 ppm (Feth and others, 1965). As much as 42 percent (475,000 acres) of the state's irrigable acreage is saline or alkaline (in 1960).

Evaporation of ground water in the many closed basins is the principal salinization mechanism in Nevada. There are two places where TDS in ground water exceeds 10,000 mg/L; these are point sources in the Tonopah Basin in Soda Spring Valley and Clayton Valley a few miles south and east of Walker Lake. Seepage flows from two springs contain TDS of 15,000 and 30,000 mg/L. Water from most aquifers is saline in the Smoke Creek Basin, the Desert Creek Basin, and the Black Rock Deserts in the lower Quinn River Basin, as well as in the Lovelock area in the lower Humboldt River basin. Long residence time of ground water in sediments containing large amounts of salts is the primary reason for TDS concentrations greater than 1,000 mg/L, and whereby mineral content increases toward the center of the basins. The areas of Carson Sink, Walker Lake, and east of Pyramid Lake contain about 50 to 60 percent of all the mineralized ground water in the state. North of Lake Mead is an area of a few thousand acres that is underlain by saline water ranging in TDS from 1,000 to 3,000 mg/L; soluble salts in the aquifers are the most likely source of the high salt content (Fuhriman and Barton, 1971).

Closed-basin lakes of high salinity include Big Soda Lake, Pyramid Lake, Walker Lake, Winnemucca Lake, Carson Sink, Rhodes marsh, and a closed-basin sump northeast of Fernly. The primary sources of salt in some of these lakes probably are unconsolidated sediments in the Lahontan Valley Group, from which highly concentrated ground water discharges into rivers and lakes (Whitehead and Feth, 1961).

Ground water in geothermal areas frequently exceeds 1,000 mg/L TDS (Thomas and Hoffman, 1988). Evaporative concentration of meteoric water before recharge and mineral-rock interactions govern the chemical composition of thermal waters from springs and wells in the Bradys Hot Springs geothermal area of Churchill County. The hottest water sampled at Bradys Hot Springs contained 2,600 mg/L but concentrations of greater than 6,000 mg/L have been reported from the Desert Peak area (Welch and Preissler, 1986).

New Hampshire: Tidal waters have intruded aquifers in the Portsmouth area (Newport, 1977).

New Jersey: Saline ground water (TDS>1,000 ppm) underlies most of the southern and southeastern parts of the state at depths at or greater than 1,000 ft below land surface. Depth to saline water decreases toward the coast in the south and is generally less than 500 ft below land surface along the entire New Jersey coastline (Feth and others, 1965). Two major water bodies of salty ground water are present along the New Jersey Atlantic Coast: a shallow one in Pleistocene deposits and a deep one, generally below the 800-ft sand. The position of the salt-water/fresh-water boundaries in those water bodies depends on the head distribution of fresh water in the respective units (Upson, 1966). Salt water

47

from tidal estuaries and bays has intruded into water table and artesian aquifers due to pumping and dredging in the areas of Sayreville (Raritan River), Gibbstown-Paulsborough, Newark (Passaic River), Rahway, Camden (Delaware River), and Salem.

New Mexico: With the exception of isolated areas in the north-central and southwestern parts of the state, saline ground water (TDS>1,000 ppm) underlies New Mexico at shallow depth (<500 ft below land surface). TDS concentrations vary considerably with local well waters exceeding 35,000 ppm in Chaves, Sierra, and Otero Counties (Feth and others, 1965). Almost all aquifers in the state contain fresh as well as saline water. Seventy-five percent of New Mexico's ground water is too saline for most uses (Ong, 1988). Approximately two-thirds of the shallow ground water in storage in the eastern Tularosa Basin is slightly saline (TDS between 1,000 and 3,000 mg/L). In the entire Tularosa Basin, possibly 180 million acre-ft of brine (TDS>35,000 mg/L) are in storage, compared to only approximately 1.4 to 2.1 million acre-ft of fresh water (Orr and Myers, 1986). More than 22 percent (191,000 acres) of New Mexico's irrigable area were considered saline or alkaline in 1960 (Fuhriman and Barton, 1971). High salinities in valley-fill aquifers are the result of recharge from saline base flow from rivers and the result of irrigation return flows (Ong, 1988). In parts of south-central New Mexico, the potential yield ratio of usable ground water to impotable ground water is smaller than 1:1,000 (Scalf and others, 1973).

Heavy pumpage has caused salt-water intrusion from deep, saline bedrock formations into producing aquifers at various places throughout the state (Newport, 1977), such as in the Roswell Basin of southeastern New Mexico. This problem is anticipated in the future for the Tularosa and Estancia ground-water basins (Smith, 1989). With the exception of recharge areas in the west, ground water in the Tularosa Basin is saline, whereby concentrations generally increase from west to east. In the Pecos River Basin, saline water from deeper strata mixes with less saline water in shallow units (Ong, 1988). Water in the Pecos River below Alamogordo Dam is slightly saline and of the calcium-sulfate type. Spring inflow in the Malaga Bend area (TDS up to 270,000 ppm) and irrigation-return flow from Roswell southward increases salinity in the river and changes the water type to a sodium-chloride type. The resulting brine discharges at the surface along the Pecos River south of Carlsbad. (Scalf and others, 1973). Ground-water withdrawal in the Rio Grande drainage basin exceeds annual recharge by far; therefore, ground water in storage is being depleted steadily, with an accompanying deterioration in quality (Kelly, 1974).

Naturally occurring brine mixes with fresh water in the shallow subsurface in Eddy County before discharging into the Pecos River. To alleviate surface-water-quality deterioration, brine was pumped from the aquifer into Northeast Reservoir (Havens and Wilkins, 1979). Brine leakage from this reservoir added chloride to the river and caused gradual chloride increases from an average of 140 tons per day between 1952 and 1963; 167 tons per day between 1963 and 1966, and 256 tons per day between 1967 and 1968 (Havens and Wilkins, 1979).

New York: Most of the central and southern parts of New York are underlain by saline ground water (TDS>1,000 ppm) at variable depths between <500 ft below land surface and >1,000 ft below land surface. Saline water appears to be shallowest along Lake Ontario and adjacent to the St. Lawrence River in northwesternmost New York (Feth and others, 1965).

Natural saline water occurs at shallow depth in western and central New York. In northwestern Cattaraugus County, brine moves from bedrock into shallow unconsolidated aquifers (U.S. Geological Survey, 1984). Oil and gas seeps have been known to occur in western New York since historic time, as documented by stories about Indians collecting oil from the Seneca Oil spring near Cuba, New York. Other sites of oil springs are at Freedom, Allegany County, and around Canandaigua Lake. This discharge of gas or oil at land surface probably is too small to have contaminated water supplies; an exception to this may be Oil Creek near Cuba. However, salt springs that issue from Silurian rocks containing halite in western New York have contaminated surface and shallow ground water. Only a few of these salt-water springs appear to be associated with oil- and-gas bearing units (Crain, 1969). Some of the Paleozoic sedimentary rocks underlying the glacial debris contain lenses of salt, contributing to naturally saline ground water in Paleozoic rocks, in the overlying glacial debris, and in local springs (Fairchild, 1935; Diment and others, 1973).

Saline water in the Jamestown area may be due to upward migration of deep saline water (from 1,500 to 2,000 ft below mean sea level) along natural breaks in the rocks and/or along abandoned oil and gas wells that have not been adequately plugged. Salt beds and highly mineralized water underlie areas of Chemung County in western New York. Old abandoned gas wells are conduits for the upward migration of these saline waters. Fresh-water aquifers in the area generally contain less than 10 mg/L; wells contaminated by brines from deeper strata contain between 100 and 500 mg/L. Similar conditions may exist in the Susquehanna River Basin (Gass and others, 1977).

North Carolina: Saline ground water containing TDS concentrations greater than 1,000 ppm occur at depths less than 500 ft along the Atlantic Coast. The depth to saline water generally increases inland (Feth and others, 1965). The three major aquifers of northeast North Carolina all contain salt water in their eastern portions (Wilder and others, 1978), imposing a threat if heavy pumpage of fresh water should occur.

North Dakota: The entire state is underlain by saline ground water (TDS>1,000 ppm) at depths less than 500 ft below land surface. Concentrations of greater than 10,000 ppm occur in the northeastern portion of the state and in northern Burke County (Feth and others, 1965).

Artesian pressure on deep saline aquifers and saline seep are major salinization hazards. In the Red River Valley, heads of deep saline aquifers are above land surface, resulting in the surface discharge of saline water (Atkinson and others, 1986). In addition, pumping of fresh ground water from overlying aquifers has induced upward migration of saline ground water from deep aquifers in the Red River Valley (Newport, 1977).

Closed basins, such as the Devils Lake chain, contain slightly saline water or brine. The total surface area of Devils Lake has decreased from 90,000 acres in 1867 to 6,500 acres in 1940, transforming the

49

water to a shallow body of stagnant brine. Salt concentrations in the lake range from 6,000 to 25,000 ppm. In the East Stump Lake, TDS ranges from 19,000 to 106,000 ppm (records from 1889–1923 and 1948–1952) (Krieger and others, 1957). Maximum TDS concentrations in the states aquifers are: Glacial deposits, 5,000 mg/L; Fort Union, 7,000 mg/L; Dakota, 11,000 mg/L; and Madison, 350,000 mg/L (Taylor, 1983).

With the exception of the eastern state-border area, all land areas of North Dakota can be considered potential saline-seep areas (Bahls and Miller, 1975). As of 1960, approximately 31 percent (817,000 acres) of the state's irrigable land was considered saline or alkaline (Fuhriman and Barton, 1971).

Ohio: The entire state of Ohio is underlain by saline ground water (TDS>1,000 ppm) at depths less than 500 ft below land surface (Feth and others, 1965). Nevertheless, upconing and intrusion of natural salt water has not yet been a major problem (Atkinson and others, 1986).

Oklahoma: Almost the entire state of Oklahoma is underlain by shallow saline ground water (TDS>1,000 ppm at depths less than 500 ft below land surface). Elevated TDS concentrations in excess of 3,000 ppm can be found in the southwestern part of the state and in spring waters in Blaine, Harper, and Alfalfa Counties (Feth and others, 1965).

Natural salinity and oil-field brines constitute problems in the Cimarron Terrace from Woods County southeast to Logan County (Atkinson and others, 1986). Salt springs and seeps issuing from underlying salt beds increase the salinity of the Cimarron River near Mocane. The salinity of the river water is increased further in the lower reaches by natural salt beds and oil-field brines (Krieger and others, 1957). Naturally occurring salt water in Permian rocks also accounts for ground-water salinization in an area near Dover, Kingfisher County (Oklahoma Water Resources Board, 1975). Overdevelopment of ground water has resulted in salt-water intrusion at many localities throughout the state. One example is the overdevelopment from the Garber-Wellington aquifer in central Oklahoma (Atkinson and others, 1986). Approximately 23 percent (194,000 acres) of the state's irrigable land was considered saline or alkaline in 1960 (Fuhriman and Barton, 1971).

Oregon: Local areas of shallow saline ground water (TDS>1,000 ppm at depths <500 ft below land surface) are reported for western Lewis, southeastern Columbia, southwestern Clark, western Multnoma, northwestern Clackamas, central and northern Yamhill, most of Washington, and central Harney Counties. In addition, isolated springs and wells in the western, southern, and eastern portions of the state may exhibit TDS concentrations greater than 3,000 ppm (Feth and others, 1965). Shallow saline ground water can be found in valley sites near principal streams in many areas of western Oregon underlain by sedimentary and volcanic rocks. Locally, saline ground water discharges to streams (U.S. Geological Survey, 1984).

The majority of salt-water problems due to intrusion have occurred in the northwestern part of Oregon. Upconing and/or lateral intrusion has been reported near Portland in Multnomah County and near Salem in Yamhill, Marion, and Polk Counties (Atkinson and others, 1986).

Saline lakes of Oregon include Alkali Valley, Abert Lake, Harney Lake, and Summer Lake (Hardie and Eugster, 1970).

Approximately 7 percent (103,000 acres) of the state's irrigable area was considered saline or alkaline in 1960 (van der Leeden and others, 1975).

Pennsylvania: Saline ground water containing >1,000 ppm TDS underlie the western half of the state at depths <500 ft below land surface (Feth and others, 1965).

South Carolina: Saline ground water (TDS>1,000 ppm) underlies the eastern part of the state within a belt reaching from the coast approximately 100 miles inland in the south and 30 miles inland in the north. Depths to saline water generally exceeds 500 ft with the exception of coastal areas in the southeast and the northeast, where depths to saline water are less than 500 ft (Feth and others, 1965). The zone of salt water, which extends through the center of the state from the southwest to the northeast, is associated with salt deposits (Speiran and others, 1988).

Heavy pumpage has caused upward and downward intrusion from layered saline aquifers and lateral intrusion from the Atlantic Ocean in the Beauford and Charleston areas (Miller and others, 1977). It has lowered water levels in the Black Creek aquifer to more than 100 ft below land surface, threatening fresh ground-water sources with the potential of lateral and upward intrusion of saline water (U.S. Geological Survey, 1984).

Background concentrations of sodium and chloride in the Black Creek aquifer of Horry and Georgetown Counties are less than 280 mg/L and less than 40 mg/L, respectively. Concentrations greater than those may indicate mixing of sea water and fresh water. There is no indication of sea-water intrusion caused by pumping, suggesting that deterioration of water quality is associated with vertical migration of salt water. Residual sea water may not be flushed from downdip portions of aquifers, resulting in a fairly wide zone of dispersion and diffusion (Zack and Roberts, 1988).

South Dakota: Ground water of highly variable quality containing between 1,000 and 10,000 ppm TDS underlies the state at depths between less than 500 ft and greater than 1,000 ft below land surface (Feth and others, 1965).

Upconing of saline water due to pumpage of fresh water aquifers has been reported at various places throughout the state, including in the Black Hills area of southwestern South Dakota (Newport, 1977). But the major salt-water problem in the state is associated with the occurrence of saline seeps, which exist throughout most of the northern, central, and northeastern parts of the state (Atkinson and others, 1986).

In 1950 it was estimated that approximately 12,000 to 15,000 artesian wells within the state leak water into aquifers above them. Inadequately plugged test holes drilled for oil, gas, and uranium may permit upward migration of saline water even in areas where no production is occurring. Maximum TDS concentrations in the state's aquifers are: glacial deposits, 10,000 mg/L; Dakota, 8,000 mg/L; Inyan Kara,

10,000 mg/L; Sundance, 7,600 mg/L; Minnelusa, 4,300 mg/L; Madison, 120,000 mg/L; Red River, 130,000 mg/L; and Deadwood, 40,000 mg/L (Taylor, 1983).

More than 70 percent (1,196,000 acres) of the state's irrigable land surface was considered saline or alkaline in 1960 (Fuhriman and Barton, 1971).

Tennessee: The western two-thirds of the state are underlain by ground water containing more than 1,000 ppm TDS. The depths to the interface between saline and fresh water is <500 ft in the center of the state but generally more than 1,000 ft in the western part of the state. In addition, a large area of extremely saline water, with TDS between 3,000 ppm and 35,000 ppm, underlies the midsection of the state at depths less than 500 ft below land surface (Feth and others, 1965). No salt-water problems have been reported in Tennessee (Atkinson and others, 1986).

Texas: Much of Texas is underlain by saline ground water (TDS>1,000 ppm) at depths <500 ft below land surface. Depth to saline water is somewhat greater in the southern part of West Texas, in south-central and east-central Texas, in East Texas, and along the Gulf Coast. The occurrence and availability of subsurface saline water in Texas were summarized for all major aquifers by Snyder and others (1972). Included in their listings are waters with TDS greater than 3,000 ppm. Brine springs and shallow saline ground water with TDS concentrations in excess of 3,000 ppm occur in the Rolling Plains of north-central Texas. Other high-saline areas are along the Pecos River and the Rio Grande in West Texas. Ground-water quality in the Salt Basin of Trans-Pecos Texas deteriorates in a northward direction, with TDS ranging from 1,550 mg/L to more than 6,000 mg/L in the heavily pumped areas of the Dell City area (Davis and Gordon, 1970). High-saline areas also occur locally throughout South Texas (Feth and others, 1965). In Central Texas, a zone of saline water borders the Edwards aquifer, which is a major source of fresh water in the area, to the east of the so-called "bad-water line" (Senger and others, 1990).

Leakage of salt water and gas from fault zones has been detected in Wood County. As a direct result of heavy pumpage salt-water intrusion has occurred in the Wintergarden area southwest of San Antonio, and in East Texas along the Gulf Coast.

Tributaries of the Red River contain high salt loads. The Salt Fork of the Red River and Mulberry Creek contain high proportions of calcium sulfate derived from gypsum. The Prairie Dog Town Fork of the Red River is highly saline with common salt and the Pease River has high sulfate and chloride concentrations. The low flows of the Salt Fork and the Double Mountain Fork of the Brazos River are at times slightly to moderately saline. Saline surface water is most common in western and northwestern Texas where rainfall is low, evaporation is high, and where rock formations at the surface contain large amounts of readily soluble minerals. Salt springs contribute to the salinity of the Colorado River above Colorado City. Saline surface waters in the eastern and southern parts of the state often originate in the west or northwest or are due to oil-field brine pollution. Along the Gulf Coast sea water mixes with surface water in the tidal reaches of the rivers, such as in the Calcasieu River channel, Calcasieu Lake, Sabine Lake, and Lower Sabine River (Krieger and others, 1957).

The Ogallala aquifer is contaminated by salt water downstream from every one of about 30 large pluvial lake basins that contain saline water or saline lacustrine sediments. In some instances, saline ground water may cover an area of several hundreds of square miles (Reeves, 1970). Saline ground water and brine lakes occur in parts of Terry and Lynn Counties. The source of the salt water is concentrated brine which exists in the Tahoka, Ogallala, and Cretaceous rocks of the area. Irrigation pumpage caused a migration of the interface between fresh water and salt water, resulting in salinization of water wells and soils in topographically low areas where evaporation of water-logged soils occurs. Chloride as well as sulfate concentrations in closed drainage basins of the area range as high as 170,000 mg/L. Disposal of brine and brine spills at a sodium-sulfate solution mining operation has contributed locally to the salinization of soils and ground water, but is not believed to play a major role as salinization mechanism (Bluntzer, 1982).

Localized chloride contamination of Pliocene sands in Texas may be caused by casing corrosion of active or inactive water wells, permitting flow of highly mineralized water from shallow strata to deeper aquifers (Sayre, 1937, in Gass and others, 1977). Serious regional contamination problems may have occurred in areas where highly pressured brine aquifers occur, such as the Rustler Formation of southwest Texas, the Coleman Junction Limestone in west-central Texas, and the deep Miocene brine aquifers of the Gulf Coast. Plugging of boreholes often was done by simply putting wood, mud, or rocks into the hole and dry holes were often left uncased (McMillion, 1965).

East Texas towns, pumping from the Woodbine Formation more than 20 miles from the outcrop, often experience TDS of 1,500–4,000 ppm in their drinking-water supplies. For example, water samples from the town of Anna, Collin County, contained 4,112 ppm TDS (Sundstrom and others, 1948). The source of saline water may be intrusion of brine from the Tyler Basin onto the North Texas Shelf (Parker, 1969).

Heavy pumpage of irrigation water has caused intrusion of highly mineralized river water into ground water in the Pecos River Valley, especially in Reeves and Pecos Counties. By far ground-water withdrawal in the Rio Grande drainage basin exceeds annual recharge; therefore, ground water in storage is being depleted steadily with accompanying deterioration in quality (Kelly, 1974). The same mechanism of overpumpage is responsible for salt-water intrusion into the Beaumont clay aquifer in Brazoria County, the Chicot aquifer in Orange County, the Seymour Formation in Knox County, and into alluvium in Ward County (Scalf and others, 1973). Winslow and others (1957) estimated that the interface between fresh and saline ground water in Harris County is moving toward the centers of heavy pumping at a speed of a few hundred feet per year.

Utah: Salinization in the state can in part be attributed to vertical movement of saline water from saline aquifers (Waddell and Maxell, 1988). Many areas in the western and southeastern parts of the state are underlain by slightly to very saline water at depths less than 500 ft below land surface (Feth and others, 1965). Extensive areas of low-quality ground water can be found in Uintah, Emery, and Grand Counties, and in the Price, San Rafael, and the Dirty Devil River basins. Saline ground water in the Bear River Basin may be caused by (a) long residence time of water in fine-grained deposits containing soluble salts, (b) evapotranspiration, (c) recharge with saline irrigation-return water, and/or (d) lateral movement of saline water from areas of large thermal springs. Highly mineralized ground water also underlies areas west of Great Salt Lake (Fuhriman and Barton, 1971). Highly saline water can be found in closed-basin lakes and their tributaries, such as Bonneville Salt Flat, El Monte Hot Spring, Great Salt Lake, Hooper Hot Spring, Locomotive Spring, Stinking Spring west of Corinne, and Utah Hot Springs. Bedded halite and gypsum underlying the Sevier River Basin contribute to the high content of dissolved solids carried by the Sevier River (Whitehead and Feth, 1961). Mineral springs contribute highly mineralized water to the Virgin River below La Verkin (Krieger and others, 1957). Although these springs contribute only approximately ten percent of the water in the Virgin River, contribution of salt amounts to 60 percent of the river's total salt load (Thorne and Peterson, 1967).

In the western part of the state and in the Great Salt Lake area the potential of pumpage-induced intrusion of salt water is high (Newport, 1977). Recharge from precipitation in the Northern Great Salt Lake Desert is approximately 200 times higher than surface outflow to the Great Salt Lake (Stephens, 1974). High evaporation rates in this area account for briny conditions in the shallow aquifer composed of salt and lake-bed deposits. A second aquifer in the area, which is made up of surficial and buried alluvial fans along mountain flanks, yields fresh to moderately saline water. However, the most extensive aquifer underlying the area yields brine from 1,000 to 1,600 ft below land surface in the Bonneville Salt Flats area (Stephens, 1974). Of all the shallow, recoverable ground water in the area, approximately 75 percent are highly saline or briny (Stephens, 1974).

Streams that contribute to the salinity of the Colorado River include the Duchesne, Price, San Rafael, Dirty Devil, Escalante, Paria, and Virgin Rivers. The principal sources of the salt content are irrigation-return waters, natural runoff from soils developed on shale, and saline springs discharging from the shale and other salt-bearing formations (U.S. Geological Survey, 1984).

Mineral and thermal springs occur in fault zones along the Wasatch mountain range, along the Hansel Valley fault in northwestern Utah, on the east side of Promontory Point, near Howell, Grantsville, Callao, and Gandy, near Timpie at the northwestern tip of Stansbury Mountains, and in Skull Valley, (Fuhriman and Barton, 1971).

Virginia: Saline ground water at depths less than 500 ft below land surface can be found along the Atlantic Coast in southeastern Virginia. Depth to saline water increases to greater than 1,000 ft below land surface in the northeastern coastal area and in the westernmost parts of the state (Feth and others, 1965).

Historic changes in chloride content of water wells are local rather than regional phenomenons. In general, a wedge of salt water coincident with the mouth of Chesapeake Bay extends into the York-James and southern Middle Neck Peninsula, where the greatest chloride increase (175 mg/L) was measured. Possible sources of salt water in ground water of the Virginia Coastal Plain are natural or induced sea-water intrusion, incomplete flushing of marine sediments, solution of evaporite deposits, or concentration by movement of water through clay-rich sediments (Larson, 1981).

Washington: Some of the coastal saline waters in the state are probably relict sea water or connate water (van der Leeden and others, 1975). Saline lakes in the state include Hot Lake, Lake Lenore, and Soap Lake (Hardie and Eugster, 1970). Of the total irrigable land area in the state, approximately 12 percent (266,000 acres) was considered saline or alkaline in 1960 (Fuhriman and Barton, 1971).

West Virginia: With the exception of the northeasternmost part of the state, West Virginia is underlain by saline water at depths ranging from less than 500 ft below land surface in the west to northwest to greater than 1,000 ft in the south and east (Feth and others, 1965). Salt-water contamination has not been reported in the state but there is a potential for movement of residual saline water toward pumping centers (Newport, 1977).

Historic records report salt and oil springs and shallow brine occurrences at various localities in West Virginia. Among these are Campbells Creek, Kanawha County, and Bulltown, Braxton County. At some localities, salt water occurs significantly higher than fresh water, suggesting salt-water movement upward through uncased holes and laterally into fresh-water aquifers. The town of Walton, Roane County, was without potable water supplies for an extended period of time due to salt-water contamination (Bain, 1970).

Natural salt springs in the Kanawha River valley occur near the mouth of Campbells Creek at Malden, Kanawha County. In most areas of the western half of the state, salty ground water below major stream valleys occurs at depths approximately 100 to 300 ft below land surface. Ground water turns from salty to brine (Cl>30,000 mg/L) at depths of 1,000 ft to 5,000 ft. In the last few decades, salt-water migration toward the land surface has been caused by vertical leakage along hundreds of unplugged wells and test holes. These wells had been drilled during exploration for brine, oil, gas, and coal, and commonly were abandoned uncased or improperly plugged. In Fayette County, chloride concentration of ground water increased from 53 mg/L to greater than 1,900 mg/L within 5[1]/2 years due to fresh-water pumpage and inflow of salt water from abandoned boreholes. Heavy pumpage of fresh ground water from bedrock aquifers at Charleston caused salt-water intrusion and chloride increases from less than 100 mg/L to greater than 1,000 mg/L (Wilmoth, 1972).

Wisconsin: Saline ground water occurs at depths less than 500 ft below land surface along the shore of Lake Superior and in some areas along the shore of Lake Michigan (Feth and others, 1965). Heavy pumpage has caused lateral intrusion of this saline water into fresh water at various locations in the state (Newport, 1977).

Wyoming: Saline ground water underlies the state at depths ranging from less than 500 ft below land surface in the southwest, center, north, and northeast to greater than 1,000 ft below land surface in the southeast and scattered areas in the center (Feth and others, 1965)

Maximum TDS concentrations in the state's aquifers are: Wind River, 6,500 mg/L; Mesaverde, 16,300 mg/L; Frontier, 20,000 mg/L; Cloverly, 24,000 mg/L; Sundance, 4,700 mg/L; Nugget, 4,900 mg/L; Casper, 9,600 mg/L; and Tensleep, 280,000 mg/L (Taylor, 1983). The average TDS load of the Green River has increased from 410,000 to 612,000 tons per year (49 percent) at a time during which the average streamflow has increased by only 5 percent. Saline-water seeps and irrigation-return flows in the Big Sandy River account for the increase in salinity (U.S. Geological Survey, 1984).

About 22 percent (280,000 acres) of the state's irrigable land was considered saline in 1960 (Fuhriman and Barton, 1971).

3.2 Halite Solution

3.2.1. Mechanism

Many sedimentary basins in the United Sates contain thick layers of rock salt (halite) (Fig. 2). In most instances these layers occur as beds but in some cases salt has been deformed and is present now in the form of salt diapirs or salt domes. Dome provinces are found predominantly along the Gulf Coast, with more than 300 individual domes in Texas, Louisiana, Mississippi, and Alabama. The depth to salt beds and salt domes varies widely, from more than 10,000 ft below land surface in Florida to a few hundred ft below land surface in parts of Texas and Louisiana, or at land surface at places in Utah (Dunrud and Nevins, 1981). Although halite is highly soluble, some of these deposits have been stable for hundreds of millions of years with little or no dissolution going on, indicating that they are not in contact with circulating fresh water. Other salt occurrences are located within local or regional ground-water flow systems and are being dissolved, predominantly along the tops and margins, causing salinization of ground water. For example, the Canadian River and Pecos River Valleys of Texas and New Mexico overly areas where as much as 200 m of salt have been dissolved by circulating ground water (Gustavson and others, 1990). Besides this natural solution of salt, solution mining associated with the recovery of salt, oil, gas, or sulfur has been practiced in the country for more than 100 years. In 1956, salt production amounted to more than 24 million tons, with approximately 75 percent produced by salt dissolution and 25 percent produced as rock salt (Pierce and Rich, 1962). Natural and man-made removal of salt has caused subsidence and collapse where salt occurs at shallow depths.

Johnson and others (1977) summarized the requirements necessary for dissolution of halite as being (1) a supply of water unsaturated with respect to halite, (2) a deposit of salt through which or against which the water flows, (3) an outlet that will accept the resultant brine, and (4) energy (such as a hydrostatic head) to cause flow of water through the system. If those requirements are met and if the outlet of the resultant brine is at land surface or becomes part of an underground fresh-water flow system, mixing of brine and fresh water will occur. For example, such process was described by Hitchon and others (1969) in the Mackenzie River Basin of western Canada, by Gillespie and Hargadine (1981) in the Kansas Permian

Basin, and by Gustavson and others (1980) on the eastern flank of the Texas Permian Basin (Fig. 20). Discharge of the resulting brine at land surface or in the shallow subsurface at those sites has been documented by Hitchon and others (1969) in Canada, by Whittemore and Pollock (1979) in Kansas, and by Richter and Kreitler (1986a,b) in Texas. In each case, local meteoric water infiltrates the ground and dissolves halite in the shallow subsurface on its way to regional discharge areas at topographically low areas.

Discharge of halite-solution brine and further evaporation of the salt water at land surface can lead to the development of salt flats. A series of such salt flats, covered by gypsum and halite crusts, occurs in the Rolling Plains of north-central Texas and southwestern Oklahoma, as described by Ward (1961) and by Richter and Kreitler (1986a,b). Salt-water discharges at those areas (Fig. 21) ranges from a few thousand milligrams per liter to greater than 150,000 mg/L dissolved chloride, affecting surface-water quality for hundreds of miles downstream from salt-emission areas.

3.2.2. Composition of Halite and other Evaporites

Halite occurs in the subsurface in form of bedded or domal salt. Depending on the depositional history, halite deposits may be associated with other chloride salts (for example, carnallite, $KMgCl_3 \cdot 6H_2O$ or sylvite, KCl), with sulfates (for example, polyhalite, $K_2Ca_2Mg\ [SO_4]_4 \cdot H_2O$; anhydrite, $CaSO_4$; or gypsum, $CaSO_4 \cdot 2H_2O$), or with carbonates (for example dolomite, $CaMg(CO_3)_2$; or limestone, $CaCO_3$), which contribute to the overall salinity of ground water in contact with halite. Salt domes are composed mainly of pure halite, with minor amounts of anhydrite, gypsum, and limestone. Analyses of salt from the Gulf Coast salt-dome province consist of more than 92 percent NaCl with varying amounts of other mineral constituents and insoluble matter (Table 3). The insoluble or less-soluble fractions stay behind when the tops of shallow salt domes are dissolved by circulating fresh waters, forming what is known as the cap rock. Cap rock is often missing from deeper domes but can be several hundreds of ft thick in shallow domes. For example, in a mined salt dome in Harris County, Texas, limestone caprock extends to 76 ft, gypsum caprock from 76 to 107 ft, and anhydrite caprock from 107 to 1,010 ft, at which depth salt is encountered (Pierce and Rich, 1962).

Solution of halite by fresh water or dilution of halite brines produces water of relatively uniform chemical composition, as indicated by plots of chemical constituents in halite-solution samples from Texas, Kansas, and Canada (Fig. 22). Solution of mined rock salt deviates from the well-defined mixing (dilution) trends indicated on figure 22 because of its nearly pure NaCl composition. Natural solutions of halite, in contrast, often are in contact with other evaporites, such as gypsum and anhydrite, which accounts for somewhat higher concentrations of other major cations and anions, such as Ca and SO_4.

57

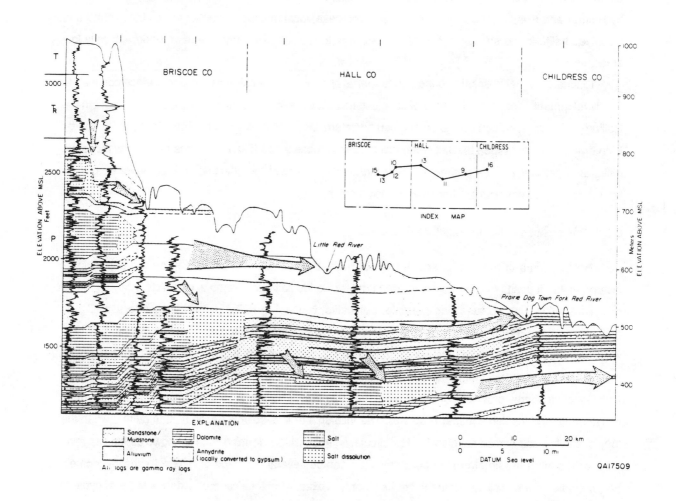

Figure 20. Solution of halite by circulating ground water, Palo Duro Basin, Texas (from Gustavson and others, 1980).

58

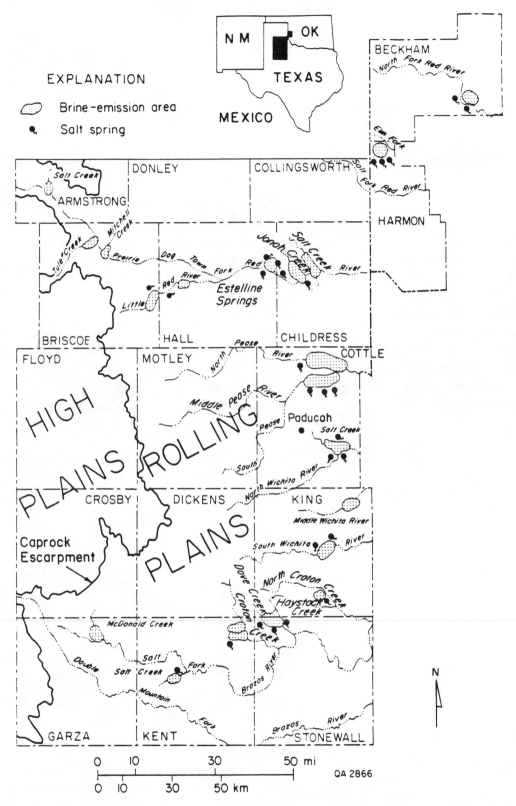

Figure 21. Discharge points of halite-solution brine at land surface in the Rolling Plains of north-central Texas and southwestern Oklahoma (from Richter and Kreitler, 1986a,b).

59

Table 3. Mineral composition of salt from domes in Louisiana and Texas (data from Pierce and Rich, 1962).

	Coastal domes, Louisiana				Interior domes, Texas	
	1	2	3	4	5	6
Sodium chloride (NaCl)	92.750	96.405	99.252	95.720	98.883	98.926
Calcium sulfate ($CaSO_4$)		0.694	0.694	3.950	1.099	1.041
Magnesium chloride($MgCl_2$)		0.012	0.012	0.008	Trace	
Magnesium carbonate ($MgCO_3$)	0.201					
Sodium carbonate (Na_2CO_3)	0.067					
Sodium sulfate (Na_2SO_4)	0.837				0.008	0.023
Calcium carbonate ($CaCO_3$)	1.804				0.010	0.100
Calcium chloride ($CaCl_2$)	0.000	0.226	0.042	0.140		
Iron and aluminum oxide (Fe_2O_3-Al_2O_3)	0.500	0.025		0.012		
Insoluble matter	3.325	0.059		0.030	Trace	

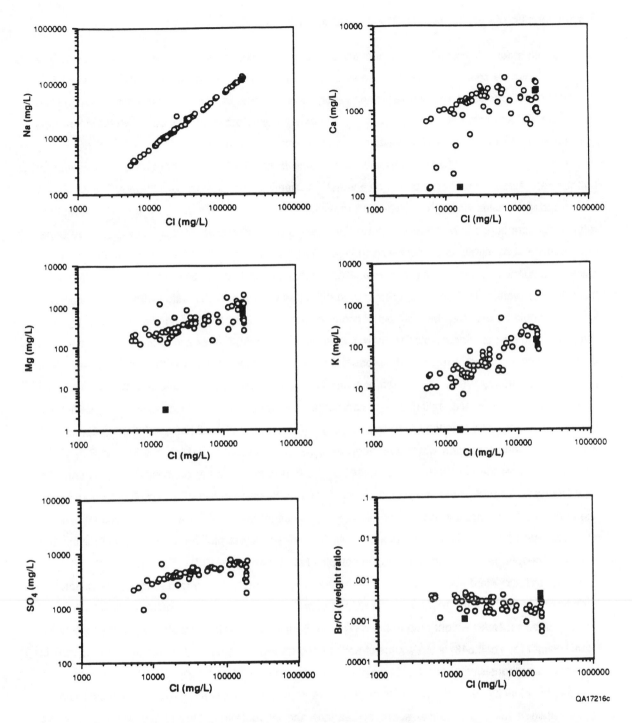

Figure 22. Bivariate plots of major ions and Br/Cl ratios versus chloride for natural halite-solution brines from Canada (data from Hitchon and others, 1969), Kansas (data from Whittemore and Pollock, 1979), and Texas (data from Dutton and others, 1985, and Richter and Kreitler, 1986a,b; open circles). Relatively little scatter in these plots suggests little variation between halite-solution brines from different areas. The composition of laboratory solutions of mined halite (solid squares; data from Whittemore and Pollock, 1979) deviates from natural solution of salt beds because of the absence of associated evaporites.

3.2.3. Examples of Geochemical Studies of Halite Solution

As with other salinization sources, identification of halite solution as source of salinity in absence of any other possible source is indicated simply by an increase in TDS and all or most other individual chemical constituents. Constituent ratios, however, have to be used to distinguish halite solution from other potential sources, such as oil-field brine. Described below are several studies where ionic ratios were used to effectively differentiate halite solution from other salt sources.

Hitchon and others (1969) determined the source of salinity in salt springs and saline seeps of northeastern Alberta to be halite, based on a variety of chemical and isotopic constituents. The K/Na ratio in salt spring samples was much lower than in typical deep-basin formation water in the area and the relative concentrations of bromide and potassium were too low to represent residual evaporative brine. Low bromide concentrations associated with high TDS contents indicate halite solution as the source of salinity, as suggested from the brine-classification system established by Rittenhouse (1967). Within this classification system, TDS contents greater than that of sea water in combination with bromide contents less than would be expected from simple concentration of sea water, suggest solution of halite (Fig. 23). Hydrogen-and oxygen-stable isotopes show a local meteoric origin, further supporting the argument that local dissolution instead of regional deep-basin discharge of formation water is the source of salt water (Hitchon and others, 1969). Dissolution of halite is also documented in southern Manitoba, Canada, where saline springs and saline seeps contribute to chloride levels of up to 600 mg/L in Lake Winnipegosis and Lake Manitoba (van Everdingen, 1971). Identification of halite solution as opposed to discharge of deep formation water was done by using the ratios of $(Ca+Mg)/(Na+K)$ and SO_4/Cl for samples with chloride concentrations greater than 10,000 mg/L. At lower concentrations, dilution with local ground water masked the salinization source. Similarly, the ratios of $(Na+K)/Cl$ and Ca/Mg were of little use in this study (apparently because of overlapping values in the endmembers). This mixing, however, was suggested from modified Piper plots, in which brackish waters plot on the theoretical mixing line between salt-spring samples and fresh runoff samples (van Everdingen, 1971) (Fig. 24).

Leonard and Ward (1962) were among the first to use the Na/Cl ratio to distinguish halite-solution brine from oil-field brine in Oklahoma. One type of brine, derived from salt springs in western Oklahoma, typically shows a Na/Cl weight ratio in the range of 0.63 to 0.65, which suggests that nearly pure halite (Na/Cl weight ratio of 0.648) is the source of sodium and chloride in those brines. Another type of brine in the same area consistently has Na/Cl weight ratios less than 0.60, whereby the ratio decreases with increase in chlorinity (Fig. 25). This type of brine was derived from oil wells. Based on this difference in ratios between the two potential sources, Leonard and Ward (1962) determined which of the two endmembers is the contaminating source in several streams of western Oklahoma, that is, halite-solution brine contributes to salinity in the Cimarron River (#5, #6), and oil-field brines contribute to salinity in the Little River (#4) and the Arkansas River (#7, #8). Surface-water degradation of streams in Kansas was

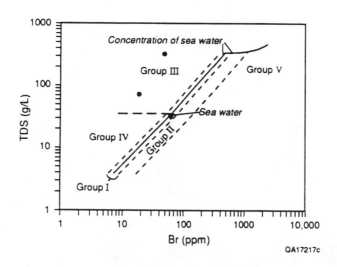

QA17217c

Figure 23. Classification of oil-field/deep-basin waters according to TDS and bromide concentrations, as suggested by Rittenhouse (1967). Brine samples from saline seeps in northeastern Alberta (solid dots) (from Hitchon and others, 1969) plot within the halite-solution group (Group III). Other groupings are: Group I, simple concentration or dilution of sea water; Group II, same as one with additional bromide, possibly added during early diagenesis; Group IV, diluted Group III waters or dilution of Groups I, II, or V with low-Br water; Group V, possibly altered bitterns left after salt deposition (from Rittenhouse, 1967).

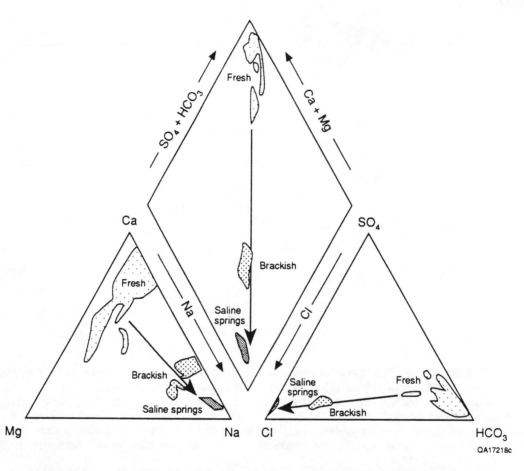

Figure 24. Modified Piper diagrams (rotated axes) of chemical composition of brine springs and surface waters in southern Manitoba. Mixing trends suggest solution of halite as the source of salinity in brackish ground water of the area (modified from van Everdingen, 1971).

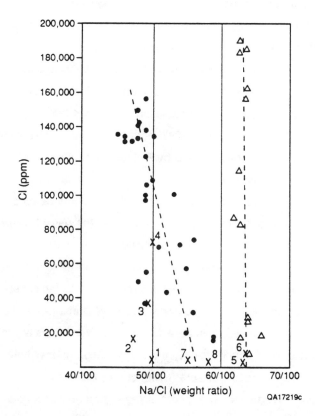

Figure 25. Comparison of Na/Cl weight ratios for oil-field brines (dots) and brines from salt springs (triangles) in western Oklahoma and southwestern Kansas (from Leonard and Ward, 1962). Weight ratios of approximately 0.65 suggest halite solution as the source of salinity in the salt springs and in some surface waters (crosses; Nos. 5 and 6), whereas ratios of less than 0.60 suggest oil-field contamination in other surface waters.

studied by Gogel (1981) using the same technique. Samples from the Ninnescah River alluvium ranged from 0.65 to 0.67 in Na/Cl weight ratios, indicating that halite solution in the underlying Wellington aquifer is the source of salinity. Measurements along Slate Creek showed the same ratio range for samples having chloride concentrations greater than 500 mg/L but a ratio of 0.91 for a sample having a chloride concentration of only 78 mg/L. This sample was obtained upstream from the saline-water inflow area and represents uncontaminated background concentrations. Very low Na/Cl ratios, ranging from 0.54 to 0.28, along Salt Creek, in contrast, suggest oil-field contamination (Gogel, 1981). The Na/Cl ratio was also used by the Oklahoma Water Resources Board (1975) to distinguish halite-solution brine from oil-field contamination in parts of Oklahoma. Weight ratios of 0.66 were indicative of halite solution in the Dover area, whereas ratios of 0.38 were indicative of oil-field pollution in the Crescent area.

Whittemore (1984) pointed out that halite solution brines in Kansas usually have lower Ca/Cl and Mg/Cl ratios and higher SO_4/Cl ratios than oil-field brines. The same relationship was found for halite-dissolution brines and deep-basin brines in Texas by Richter and Kreitler (1986a,b). However, these ratios appear to work best as tracers when little chemical reactions occur after mixing of the respective salt-water source and fresh water. To avoid this change of chemical constituent ratios by mechanisms other than mixing or dilution, Whittemore and Pollock (1979) suggested the use of minor chemical constituents, such as bromide, iodide, and lithium, that are relatively conservative in solution. Of those, bromide concentrations and Br/Cl weight ratios are the most widely used tracers because bromide is similarly conservative as chloride, and because a significant difference in Br/Cl ratios between most halite-solution brines and oil-field waters could be established (for example, Whittemore and Pollock, 1979; Whittemore, 1984, 1988; Richter and Kreitler, 1986a,b). In Kansas, halite-solution brines typically have Br/Cl ratios less than 10×10^{-4}, whereas oil-field brines typically have Br/Cl ratios greater than 10×10^{-4} (Whittemore, 1984). This difference can be used to calculate mixing curves between fresh water composition and the range of values for either endmember. By superimposing values from test holes in the Smoky Hill River area, Whittemore (1984) was able to show that halite solution is the dominant mechanism of salinization in that area (Fig. 26a). In contrast, in the Blood Orchard area south of Wichita, all of the observation-well samples indicate mixing of fresh water with oil-field brines (Fig. 26b). Note in figure 26b that the sample obtained from the Arkansas River (square) suggests halite solution; this chemical signature was derived from areas upstream from the Blood Orchard area. Also, some testhole samples plot between the two mixing fields, suggesting mixing of fresh water with both halite-solution brine and oil-field brine.

Richter and Kreitler (1986a,b) made use of the Na/Cl and Br/Cl ratios in an investigation of salt springs and shallow subsurface brines in parts of north-central Texas and southwestern Oklahoma. Using ratio ranges established by Ward (1961) for Na/Cl in Oklahoma brines and Br/Cl ratio ranges established by Whittemore and Pollock (1979) for Kansas brines, Richter and Kreitler (1986a,b) grouped 120 chemical analyses of saline waters into three groups: (1) a halite-solution group (Group A) with Na/Cl molar ratios greater than 0.95 (Na/Cl weight ratio >0.62) and Br/Cl weight ratios less than 4×10^{-4}, (2) a deep-basin

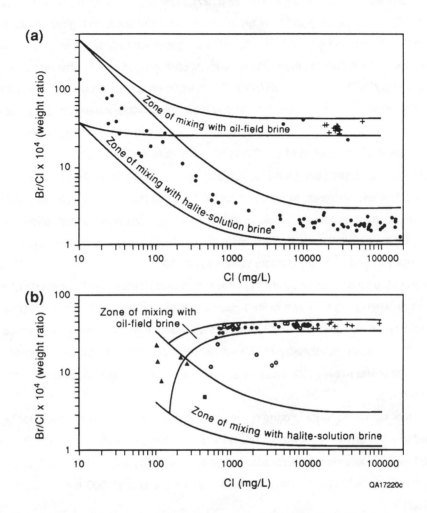

Figure 26. Differentiation of oil-field/deep-basin brine from halite-solution brine using bivariate plots of Br/Cl weight ratios versus chloride concentrations. (a) Low Br/Cl ratios typically signify halite solution (for example, Smoky Hill River, Kansas), whereas (b) high Br/Cl ratios and high TDS are characteristic of oil-field/deep-basin brines (for example, Blood Orchard area, Kansas) (from Whittemore, 1984).

group (Group C) with Na/Cl ratios less than 0.95 and Br/Cl ratios greater than 25×10^{-4}, and (3) a mixing group (Group B) with Na/Cl ratios greater than 0.95 and Br/Cl ratios greater than 4×10^{-4}, or Na/Cl ratios less than 0.95 and Br/Cl less than 25×10^{-4} (Fig. 27). After this initial grouping, other ratios, such as Ca/Cl, Mg/Cl, K/Cl, and I/Cl, supported the grouping by generally showing higher values for deep-basin brines than for halite-solution brines (Fig. 28). This suggests that all of these ratios can be used for differentiating between halite-solution brines and deep-basin brines in this area. Molar ratios of (Ca+Mg)/SO_4 and of Na/Cl are close to unity in halite-solution samples, which suggests halite and gypsum to be the main sources of dissolved constituents and distinguishes this water from deep-basin brines (Fig. 29). The third group (Group B) plotted intermediate in all these ratios. Group differences were also reflected in the isotopic composition of $\delta^{18}O$ and δD. Little difference was observed between local, fresh ground-water values and salt-spring values, suggesting dissolution of halite by local, meteoric ground water (Fig. 13). Testhole samples, in contrast, formed a trend from a light isotopic composition at low chloride concentrations to a heavy isotopic composition at high chloride concentrations, suggesting the mixing of local, fresh ground water with deep-basin brines. This mixing was also documented in the depth relationship of chloride and $\delta^{18}O$ for testhole samples (Fig. 30).

In addition to the ratios discussed above, Whittemore and Pollock (1979) pointed out the usefulness of I/Cl ratios in differentiating between halite-solution and oil-field brines. Halite-solution brines typically have I/Cl weight ratios less than 1×10^{-5}, whereas oil-field brines in Kansas have ratios greater than 2×10^{-5} (Fig. 31). A similar relationship in I/Cl ratios between halite-solution brine and deep-basin brine was found by Richter and Kreitler (1986a,b) for samples from salt springs and shallow test holes in north-central Texas (Fig. 28).

Dissolution of salt-dome halite in parts of the upper Gulf Coast of Texas accounts for high salinities in geopressured Frio Formation waters (Morton and Land, 1987). Besides the high salinity (TDS>105,000 mg/L), halite dissolution is also suggested as the source by low Br/Cl ratios (<22×10^{-4}), and by low potassium and calcium concentrations (<500 mg/L and <5,000 mg/L, respectively) (Morton and Land, 1987).

Mast (1982) suggested the use of mixing diagrams for differentiating between mixing of fresh water with naturally saline ground water and mixing of fresh water with oil-field brine in parts of Kansas. Contaminated waters were obtained from an aquifer that is in a producing oil field and is underlain by natural halite and gypsum beds from which it is separated by a shale of low permeability. Plotted on mixing graphs of SO_4 versus Na+K, SO_4 versus TDS, and SO_4 versus Cl (Fig. 32) are the endmembers of mixing, that is (1) uncontaminated fresh water, (2) naturally saline (halite-solution) water, and (3) produced (oil-field) water. Between the endmembers are drawn theoretical mixing lines that were calculated using mean values of the respective endmembers. Superimposed on the plot are all samples to be tested, allowing visual inspection of the most likely salinization source, if any, as well as an estimate of the relative percent contribution of each endmember (Fig. 32).

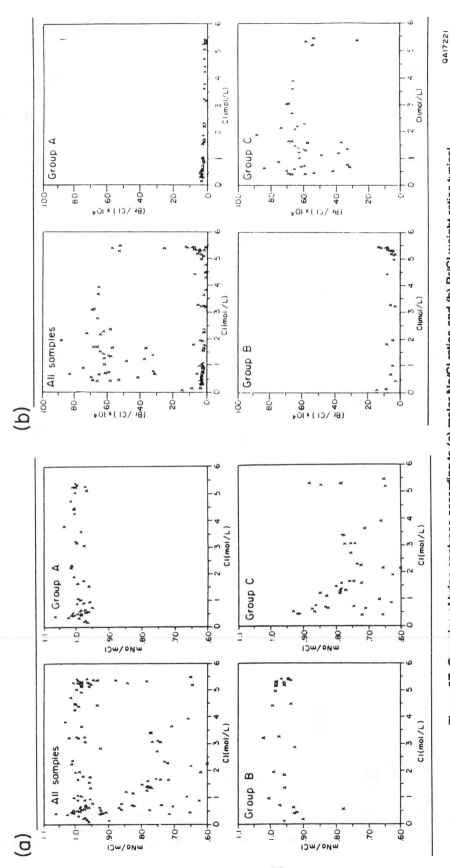

Figure 27. Grouping of brine analyses according to (a) molar Na/Cl ratios and (b) Br/Cl weight ratios typical for halite solution and oil-field/deep-basin brine. Group A: Na/Cl >0.95 and Br/Cl <25 × 10⁻⁴ (halite solution); Group B: Na/Cl >0.95 and Br/Cl >4 × 10⁻⁴, or, Na/Cl <0.95 and Br/Cl <25 × 10⁻⁴ (mixing of A and C?); Group C: Na/Cl <0.95 and Br/Cl >25 × 10⁻⁴ (deep-basin brine) (from Richter and Kreitler, 1986b).

QA17221

69

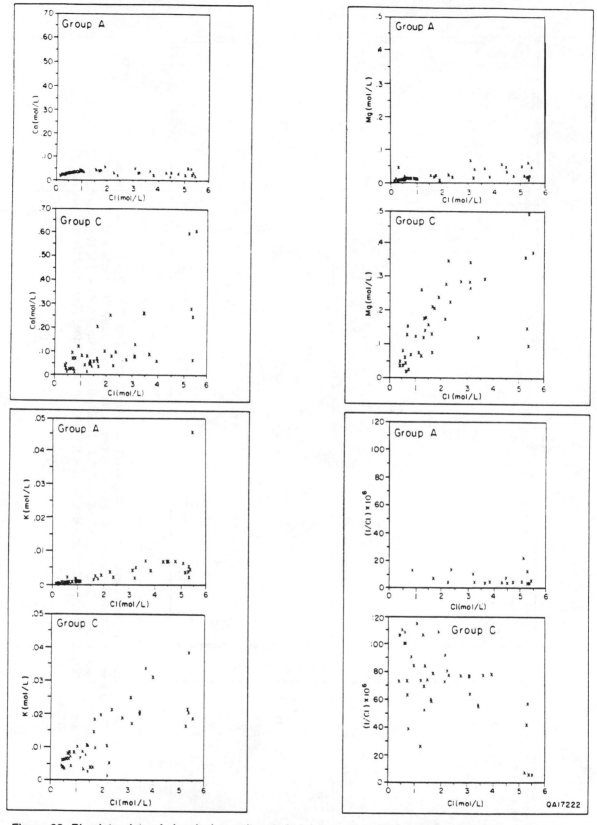

Figure 28. Bivariate plots of chemical constituents in brines collected from salt springs and shallow test holes in the Rolling Plains of Texas. Grouping of brines according to molar Na/Cl ratios and Br/Cl weight ratios (fig. 27) is supported by other chemical constituents, such as Ca, Mg, K, and I (modified from Richter and Kreitler, 1986a,b).

70

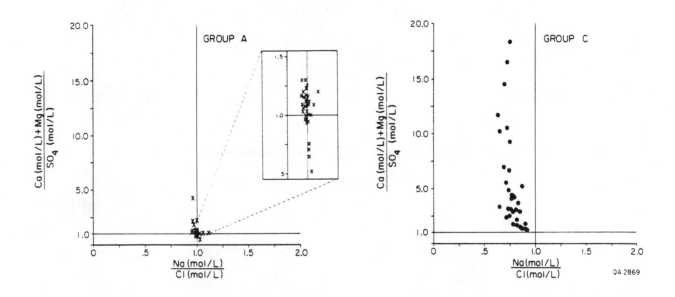

Figure 29. Molar ratios of (Ca+Mg)/SO$_4$ versus Na/Cl in salt-spring and shallow subsurface brines, Rolling Plains of Texas (from Richter and Kreitler, 1986a,b). Molar ratios close to one suggest chemical control by halite and gypsum in Group A waters. Large deviations from unity are typical for deep-basin brines and suggest such a source in Group C waters.

71

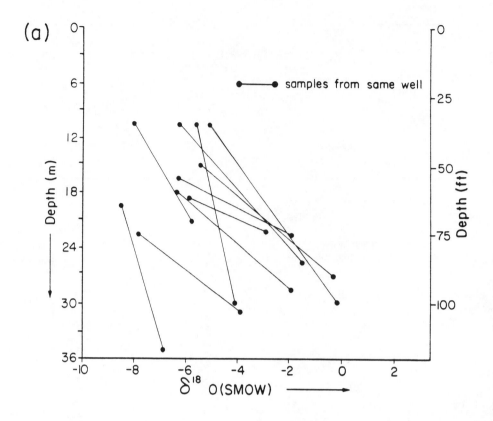

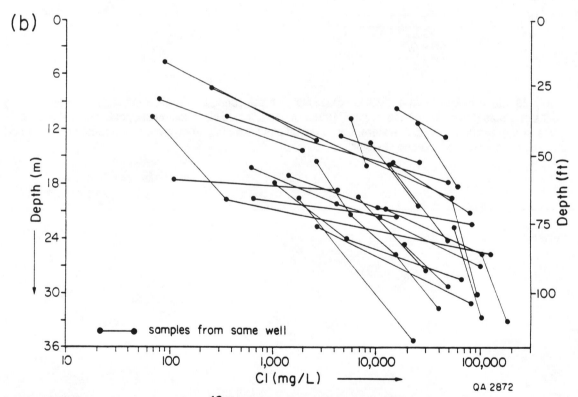

QA 2872

Figure 30. Relationship between (a) $\delta^{18}O$ and (b) Cl with depth in shallow brines in parts of the Rolling Plains, Texas. Mixing of local, meteoric fresh water with non-local brine is indicated by isotopic enrichment and by chloride increase with depth (from Richter and Kreitler, 1986b).

72

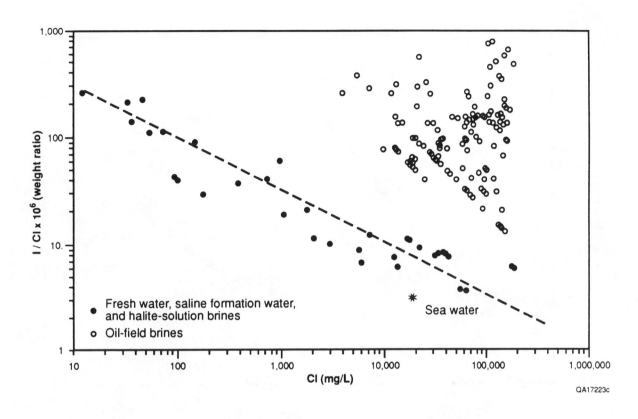

Figure 31. Differentiation between halite-solution waters and oil-field/deep-basin brines using differences in I/Cl weight ratios (from Whittemore and Pollock, 1979).

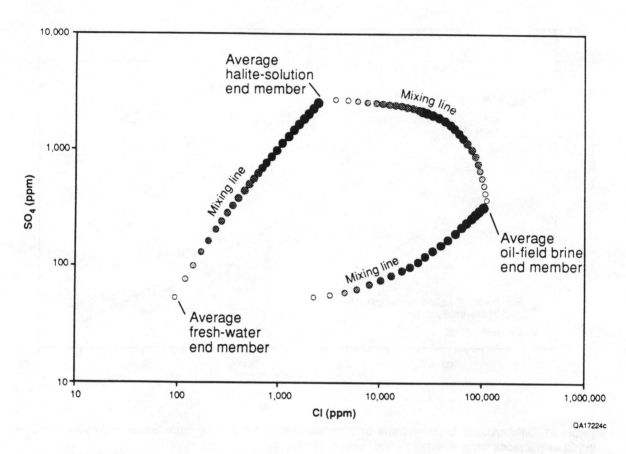

Figure 32. Theoretical mixing curves for fresh water and potential salinization sources in Kansas using sulfate and chloride concentrations (from Mast, 1982).

74

3.2.4. Significant Parameters

Halite solution produces some of the lowest Br/Cl ratios found in natural salt waters. Ratios typically are less than approximately 10×10^{-4} in halite-solution brines and greater than 10×10^{-4} in oil-field brines and formation brines (Whittemore and Pollock, 1979; Whittemore, 1984; Richter and Kreitler, 1986a,b). Ratio differences between these two potential mixing endmembers with fresh water are generally big enough to allow differentiation of the respective source in brackish water down to chloride concentrations of a few hundreds of milligrams per liter, although identification is best at high concentrations. Sea water also has a much higher Br/Cl ratio than halite-solution brine, which could allow differentiation between halite solution and sea-water intrusion in coastal salt-dome areas.

The ratio of Na/Cl works well to distinguish halite-solution brine from oil-field brine at high chloride concentrations. Sodium and chloride occur in halite at equal molar concentrations (Na/Cl molar = 1, Na/Cl weight = 0.648). Brines that originate from halite solution within a shallow ground-water flow system will exhibit a similar ratio as long as concentrations are high enough so that the Na/Cl ratio is not affected appreciably by ion exchange reactions. In most oil-field brines molar Na/Cl ratios are much less than unity. Exchange of calcium and magnesium for sodium on clay mineral surfaces and alteration of feldspar may account for the low ratios in formation brines and oil-field brines. The Na/Cl ratio is also much smaller in sea water (mNa/mCl = 0.85) than in halite-solution brine.

Ratios of I/Cl in halite-solution brines are typically small and less than oil-field/deep-basin brines, which allows separation between these two major sources of salt water.

Halite deposits are often associated with abundant beds of gypsum and anhydrite. Dissolution of these beds is reflected in molar $(Ca+Mg)/SO_4$ ratios close to unity, which is much smaller than the respective ratio in oil-field brines (>>1) and in sea water (2.3).

3.2.5 State-by-State Summary of Halite Occurrences

This section provides a state-by-state summary of some of the halite-solution occurrences in the United States (Fig. 2), as compiled from published sources.

Alabama: Bedded and diapiric occurrences of halite are known in the southwestern part of the state. All major aquifers in Marango County are affected by upward migration of brine along faults, which is probably related to salt-dome movement (Slack and Planert, 1988). Solution mining near McIntosh, Washington County, is from a salt dome where evaporites occur at 400 ft below land surface (Dunrud and Nevins, 1981).

Arizona: Large quantities of salt are discharged in the state from mineral springs and thermal mineral springs. For example, three major springs in the Mogollon Rim area of central and east-central Arizona discharge daily TDS loads of 200 tons (Fuhriman and Barton, 1971). Clifton Hot Springs discharges saline water of up to 9,000 ppm into the San Francisco River, and mineral springs near the mouths of Chevelon and Clear Creeks near Winslow contribute saline water to the Little Colorado River (Krieger and others, 1957). Five areas in the state are known to be underlain by halite, the largest of which is the Holbrook Basin in the eastern part of the state. There, depth to the top of the halite-bearing Supai Formation ranges from 650 to 800 ft below land surface (Pierce and Rich, 1962). Other areas where halite is known in the subsurface are the Red Lake and Detrital Wash areas in the western and Luke and Picacho in the south-central parts of the state. In the Red Lake and Detrital Wash areas, salt occurs at depths of 420 to 600 ft below land surface (Pierce and Rich, 1962). Top of evaporite deposits at Luke, where solution mining is practiced, is at 875 ft below land surface (Dunrud and Nevins, 1981).

Arkansas: Halite occurs in the Buckner and the Eagle Mills (Louann salt) Formations of southern Arkansas.

Colorado: Halite occurs at shallow depth in several parts of the state. Selected wells show depth to top of salt ranging from 395 ft to 6,485 ft below land surface (Pierce and Rich, 1962). An estimated 205,000 tons of salt from salt-solution enters the Dolores River of southwest Colorado annually through springs and seeps along the channel bottom (Jensen, 1978).

Florida: Halite occurs in the deep subsurface of southwesternmost Florida, at depths greater than 10,000 ft (Dunrud and Nevins, 1981).

Idaho: Halite occurs at or near land surface in the Crow Creek area (Pierce and Rich, 1962).

Kansas: Shallow Permian salt underlies most of the central and southwestern parts of the state. Salt springs, salt marshes, or direct contact with salt-bearing strata have contaminated many surface waters, such as the Smoky Hill, Arkansas, Ninnescah, Saline, and Solomon Rivers (Krieger and others, 1957; U.S. Geological Survey, 1984). Much of the saline ground water is discharged from the Wellington, Dakota, and Kiowa Formations, and the Permian Red Beds. The most widespread zone of salt is the Hutchinson salt member of the Wellington Formation. The salt bed, which is up to 350 ft thick, is easily dissolved by circulating ground water. The most damaging effects from salt dissolution are along the Smoky Hill and Solomon Rivers in eastern Saline and western Dickinson Counties and the Belle Plain area in Sumner County on the Ninnescah and Arkansas Rivers and the Slate and Salt Creeks (Gogel, 1981). Other areas contaminated by salt in Permian Red Beds are found along the lower reaches of the Rattlesnake Creek in Stafford and Reno Counties and along the South Fork Ninnescah River in eastern Pratt and western Kingman Counties. A well-documented example of chloride and sulfate intrusion from the Dakota and or Kiowa Formations is along the Smoky Hill and Saline Rivers in Russell County (Hargadine and others, 1979).

Ten minor areas of mineral intrusion occur along (1) the Solomon River in Mitchell and Cloud Counties (for example, Waconda Spring west of the city of Beloit), (2) Salt Creek in Mitchell and Lincoln counties, a tributary of the Solomon River, (3) Buffalo Creek in Jewell and Cloud Counties, (4) Salt Creek and Marsh Creek in Republican County, (5) the Cottonwood River in Marion County, (6) the Medicine Lodge River in Barber County, (7) the Salt Fork Arkansas River in Comanche and Barber Counties, (8) the

Cimarron River in Meade County, (9) the Crooked Creek in Meade County, and (10) Cow Creek in Reno and Rice Counties (Hargadine and others, 1979).

Hargadine and others (1979) studied six major areas of natural salt-water intrusion throughout the center of the state from the Oklahoma state line to the Nebraska state line. These intrusions are associated with the Dakota and/or Kiowa Formations in north-central Kansas, with Permian Red Beds in south-central Kansas, and with the Wellington Formation in east-central and south-central Kansas. Within the Wellington Formation, halite dissolution is prominent in the western part, whereas gypsum dissolution is dominant in the east.

In the area south and west of Salina, fresh water moves down through collapsed and brecciated shales to recharge the Wellington aquifer, dissolving halite within the Hutchinson Salt Member. The resulting brine then moves northward beneath the Smoky Hill River valley to the Saline River valley. Upward discharge of brine occurs through collapse structures in an area from about three miles east of Salina to just east of Solomon, where the potentiometric surface of the Wellington aquifer is above the water table in the alluvial aquifer (Gillespie and Hargadine, 1981). Between New Cambria and Solomon, approximately 369 tons of chloride enter the Smoky Hill River (J. B. Gillespie, personal communication, 1981; *in* Gogel, 1981). According to landowners, the interface between fresh water and saline water in the alluvial aquifer along the Smoky Hill River in northern Saline County is about 35 ft below land surface. Chloride concentrations of 48,000 mg/L at 94 ft depth and of 180,000 mg/L in gypsum cavities at 112 ft have been reported. A relatively impervious shale separates the shallow fresh-water unit from the deeper brine unit except for areas of collapse and brecciation caused by salt and gypsum dissolution (Gillespie and Hargadine, 1981).

Some of the brine originating from salt dissolution in Kansas does not discharge in Kansas but discharges instead in neighboring Missouri to the east. According to Banner and others (1989), regional ground-water flow that starts with meteoric recharge in Colorado picks up high loads of dissolved solids from halite-bearing formations in the subsurface of Kansas and finally discharges in central Missouri.

Most of the old solution-mining operations that were started as early as the 1880's have been abandoned. This includes the operations near Great Bend and Pawneee Rock in Barton County, near Anthony in Harper County, near Kingman in Kingman County, near Nickerson in Reno County, near Sterling in Rice County, and near Wellington in Sumner County (Dunrud and Nevins, 1981).

Louisiana: The entire state is underlain by halite with major salt-dome provinces along the coast and in the northwest. Some of the salt domes are close to land surface, such as Weeks Island, Jefferson Island, and Avery Island, where tops of salt are at 97, 69, and 15 ft below land surface, respectively (Pierce and Rich, 1962). All three domes are mined for salt. Bedded halite generally does not occur within approximately 5,000 ft below land surface (Dunrud and Nevins, 1981). Solution mining has occurred in the past in the northwestern dome province but is now restricted to domes along the Gulf Coast. Top of evaporite deposits in mined domes ranges from few ft to 6,000 ft below land surface, with most production going on since the early 1900's (Dunrud and Nevins, 1981).

Michigan: Some of the oldest salt deposits (Silurian) in the country underlie the Michigan peninsula. Combined salt thickness ranges from 1,800 ft in the basin center, where the salt is approximately 6,000 ft deep, to 500 ft at the basin margins, where the salt may be found at 500 ft below land surface (Pierce and Rich, 1962). Old and modern solution mining is concentrated along the basin margins. According to Pierce and Rich (1962), salt production in 1955 amounted to more than 5 million short tons, most of which was produced by solution mining of salt. Rock salt production in 1953 was approximately 1 million tons, mined from a salt bed 1,020 ft deep in Wayne County, near Detroit. Natural brines are produced from rocks of the Detroit River Group at Manistee and Ludington in the western part of the state, and at Midland, St. Louis, and Mayville in the eastern part of the state (Sorensen and Segall, 1973). Most of the current solution operations are at depths greater than 1,300 ft below land surface and have been going on for approximately 100 years (Dunrud and Nevins, 1981).

In Manistee County, the combined production of salt from salt beds in 1898 was approximately 435 million pounds (Childs, 1970). Production is from salt beds in the top of the Detroit River Group between 1,900 and 2,050 ft below land surface. Another producing horizon is at 3,700 ft. Modern operations require 600,000 to 800,000 gal/day of water to run economically. Ground water in and around the salt plants is contaminated with chlorides and is not a source of potable water (Childs, 1970). Backflush fluids from salt galleries, which were dumped into open pits, are part of the problem. Since the 1980's salt brines have also been continuously disposed of into Lake Manistee (Childs, 1970).

Mississippi: All of the southern part of the state is underlain by salt bed and salt domes. Solution mining has been practiced at Petal, Forrest County, and Richton, Perry County, but these operations have been abandoned (Dunrud and Nevins, 1981).

Montana: Salt beds within the Williston Basin of northeastern Montana generally occur at depths greater than 4,000 ft below land surface (Pierce and Rich, 1962).

Nebraska: Halite beds occur in southwestern Nebraska at depths greater than 3,200 ft in parts of the Northern Denver Basin (Pierce and Rich, 1962; Dunrud and Nevins, 1981).

Nevada: In the southeasternmost part of the state some domelike occurrences of salt are found in the Muddy Creek Formation along the Virgin River. These occurrences used to be exposed at land surface but are now covered by the Overton arm of Lake Mead (Pierce and Rich, 1962).

New Mexico: Halite underlies the state along its eastern state line, where salt thickness in the Permian Basin is greatest (up to 2,800 ft aggregate thickness) (Pierce and Rich, 1962). Dissolution of shallow salt on this western edge of the Permian Basin causes salinization of the Pecos River and ground water in that portion of the state. At Malaga Bend, inflow of spring water to the Pecos River contains up to 270,000 ppm of TDS. Salt occurs in the Guadalupe and Ochoa series, where it is associated with gypsum and anhydrite. In southeastern New Mexico, potassium salt with an average thickness of 4 to 5 ft are included in salt within the Salado Formation (Pierce and Rich, 1962).

New York: Halite underlies most of the western part of the state with a maximum aggregate salt thickness of 800 ft just southeast of Seneca Lake (Pierce and Rich, 1962). From there, salt thins in all directions. Associated with the salt are traces of sylvite, polyhalite, and carnallite, all potassium salts (Alling, 1928). Salt beds and highly mineralized water underlie areas of Chemung County (Gass and others, 1977) and Cattaraugus County (U.S. Geological Survey, 1984), and salt-water springs are common throughout western New York in the outcrop areas of Silurian rocks, which contain halite deposits in the subsurface. According to Crain (1969), most of these springs are not associated with deep formation waters or oil-field brines. Brines from deeper saline sources may also contaminate ground water in the Susquehanna River Basin (Gass and others, 1977). Most of the solution-mining operations in the state, starting as early as in the 1880's, have been abandoned. According to Dunrud and Nevins (1981), mining continues at Tully, Onandaga County, at Watkins Glen, Schuyler County, at Ithaca, Thomkins County, and at Dale and Silver Springs, Wyoming County. Depths to evaporite strata in those mining areas range from 1,200 to 2,500 ft (Dunrud and Nevins, 1981). In the Jamestown area, halite beds at 1,500 to 2,000 ft below land surface are penetrated by numerous abandoned gas wells, which are suspected to allow vertical migration of salt water into shallow fresh-water aquifers (Crain, 1966; Miller and others, 1974).

North Dakota: The western part of the state that is known as the Williston Basin is underlain by halite at depths greater than 3,600 ft below land surface (Pierce and Rich, 1962). Dunrud and Nevins (1981) list one solution-mining operation in the state, located near Williston in Williams County, where the top to evaporite deposits is 8,250 ft.

Ohio: Halite deposits are thickest along the eastern state line, where the combined thickness of all beds approaches 300 ft (Pierce and Rich, 1962). Salt deposits thin toward the west. Depth to salt is approximately 1,000 ft below land surface, with solution mining having been practiced at several locations. According to Dunrud and Nevins (1981), only two of the original seven mining operations are still active in the state. Those active operations are located near Akron in Summit County and near Rittman in Wayne County. Depth to evaporites in those areas is approximately 2,700 ft (Dunrud and Nevins, 1981).

Oklahoma: All of the west-central and northwestern parts of the state, including the Oklahoma Panhandle, are underlain by halite, gypsum, and anhydrite. Halite dissolution is evidenced in salt springs and in shallow saline ground water in those areas, as documented in areas such as the Cimarron River near Mocane (Krieger and others, 1957), Perkins (Leonard and Ward, 1962), or Dover in Kingfisher County (Oklahoma Water Resources Board, 1975). Halite lenses in the Flowerpot Shale within 600 ft of land surface have in the past been dissolved at most places by percolating ground water, but occasionally salt deposits still exist, such as those in the Little and Big Salt Plains at the northwestern end of the Cimarron Terrace (Oklahoma Water Resources Board, 1975).

Solution mining at 1,500 ft below land surface has been practiced on and off at Sayre, Beckham County. Pumping of salt water, from a producing interval of 12 to 558 ft at the site of a former brine spring, is practiced in Harmon County. Richter and Kreitler (1986a,b) reported a chloride concentration of 58,000 mg/L for the produced brine, which is only about 30 percent of the chloride concentration measured in two natural brine springs in the area.

Pennsylvania: All of the western and northern parts of the state are underlain by halite at depths greater than 1,000 ft. Salt beds are up to 200 ft thick, with aggregate thickness of up to 650 ft, along the western state line, but thin toward the east (Pierce and Rich, 1962).

Texas: Many areas of the state are underlain by halite deposits, with major dome provinces along the Gulf Coast, in northeast Texas, and in South Texas. Solution mining occurs predominantly in those dome provinces where depths to salt are relatively small. For example, depth to salt in Brooks Dome, Smith County, is 220 ft; in Grand Saline Dome, Van Zandt County, 212 ft; and in Palestine Dome, Anderson County, only 140 ft below land surface (Pierce and Rich, 1962). According to Dunrud and Nevins (1981), solution mining is still practiced at 12 sites in the state, ten of which produce from salt domes in Brazoria, Chambers, Duval, Fort Bend, Harris, Jefferson, Matagorda (all Gulf Coast), and Van Zandt (northeast Texas) Counties, and two produce from salt beds in Ward and Yoakum Counties.

Dissolution of salt under the High Plains Escarpment causes brine discharge in the Rolling Plains of north-central Texas and degrades surface-water quality in the Brazos and the Red River for hundreds of miles downstream (Richter and Kreitler, 1986a,b). Other affected surface streams in the area include the Pease and Wichita Rivers, the Double Mountain Fork of the Brazos, and the Prairie Dog Town Fork, and Salt Fork of the Red River. More than 3,500 tons of sodium chloride are discharged from salt seeps and springs to these rivers every day (Gustavson, 1979). Some tributaries to these rivers originate in salt flats, such as at Dove Creek and Haystack Creek in King County, Short Croton and Hot Springs in Kent County, and Jonah Creek in Childress County (Richter and Kreitler, 1986a). Gypsum and anhydrite beds are generally associated with halite beds, contributing to the overall salinization of ground water in those areas. At least some of the high salt content in the Pecos River of West Texas is due to discharge of halite-solution brine from Permian formations in New Mexico.

Mining of salt domes for commercial use (for example, brine, oil, gas, and waste disposal) has been and is being done in many salt domes along the Gulf Coast and in East Texas. Locally, these activities have caused subsidence and formation of sinkholes, such as at Boling Dome and Orchard Dome along the Gulf Coast (Mullican, 1988) and at Palestine Salt Dome in East Texas (Fogg and Kreitler (1980). At Palestine Dome, ground water recharges at highland areas, moves downward to the dome, dissolves halite, and discharges in topographic lows, such as a lake and nearby sinkholes (Fogg and Kreitler, 1980). Saline ground water at other domes, such as southeast of Mount Sylvan Dome and east of Whitehouse Dome and Bullard Dome in East Texas, may be associated with processes from the geologic past, such as previous salt-dome solution, or they may represent depositional, marine waters but not current solution of

80

domal halite (Fogg, 1980). Under certain conditions, such as at Oakwood Dome, East Texas, dissolution of salt may be slow due to a protective layer of caprock that prevents circulating water from contact with halite and low permeabilities that prevent significant ground-water circulation and with it significant removal of salt (Fogg and others, 1980). Ground-water salinity caused by natural mixing of fresh ground water with halite-solution water may be enhanced by drilling and brine disposal activities, as suggested by Hamlin and others (1988) for the lower Chicot aquifer south and west of Barber Hill Salt Dome (Chambers County). Since 1956, an estimated 1.5 billion barrels of brine have been injected into porous zones into this dome's caprock, with the present rate being 1 to 5 million barrels per month (Hamlin and others, 1988).

Utah: The Sevier River Basin in central Utah is underlain by halite and gypsum, which have contributed to the high TDS content in the Sevier River (Whitehead and Feth, 1961). Near Redmond salt is mined in open pits from the Arapien shale in the Sevier River Valley (Pierce and Rich, 1962). In the southeastern part of the state, in the Paradox Basin, depth to top of salt in selected wells ranges from 883 to 12,200 ft below land surface. There, sylvite and carnallite are associated with some of the halite deposits (Pierce and Rich, 1962).

Virginia: A shallow occurrence of salt exists in southwestern Virginia. Pierce and Rich (1962) specify a depth below land surface of 800 to 2,000 ft for this occurrence, whereas Dunrud and Nevins (1981) specify a depth of 165 to 3,300 ft. Between 1895 and 1972, solution mining, which caused local subsidence, was practiced at this shallow occurrence of salt (Dunrud and Nevins, 1981).

West Virginia: Halite that underlies the northern part of the state is generally restricted to depths greater than 5,000 ft below land surface (Pierce and Rich, 1962). Nevertheless, salt is mined at four sites located near Moundsville and Natrium in Marshall County, near Bens Run in Tyler County, and near New Martinsville in Wetzel County (Dunrud and Nevins, 1981). All of these operations are relatively recent (post-1942), which is in contrast to other, mostly much older, salt-mining operations in the country.

Wyoming: Halite occurs in the subsurface in the northeast (Powder River Basin), in the southeast (Northern Denver Basin) and in the southwest. Near the Idaho/Utah border, salt was encountered in one well at 125 ft below land surface (Pierce and Rich, 1962). In the other areas, depth to top of salt is much greater, being more than 2,700 ft in the Northern Denver Basin and more than 6,000 ft in the Powder River Basin (Dunrud and Nevins, 1981).

3.3 Sea-Water Intrusion

3.3.1. Mechanism

Sea-water intrusion has been reported from almost all coastal states in the country. In many instances, ground-water contamination from sea-water intrusion has forced abandonment of water wells or entire well fields, and much money has been spent to prevent further intrusion. Florida is the most seriously affected state, followed by California, Texas, and New York (U.S. Environmental Protection Agency, 1973).

Under natural conditions, ground-water flow in unconfined and shallow-confined coastal aquifers is toward the oceans because of flow potentials driven predominantly by topography. Generally this flow has caused flushing of saline water from coastal aquifers that may have occupied the aquifer since deposition since the last high stand of sea level, or since the last major storm flooded the coast. For example, lenses of fresh water extend as far as 75 miles from shore or to the edge of the shelf off the Florida Peninsula (Manheim and Paull, 1981). These lenses, which shield coastal ground water from saline encroachment, originate from fresh-water recharge during the exposure of the continental shelf during the Pleistocene glacial maximum (Manheim, 1990). Along the Gulf Coast of southwestern Louisiana, saline water was flushed gulfward during the Pleistocene, but moved landward through highly transmissive zones as sea level rose again (Nyman, 1978). The degree of flushing depends on many factors, such as the amount of recharge, the hydraulic head, and the permeability of the aquifer. For example, in the Houston area of Texas, fresh water has flushed the original salt water out of the aquifer to a depth of more than 2,200 ft, whereas just 50 miles toward the coast in the Galveston area salt water in the shallow aquifer has been flushed out to a depth of only 150 ft (Jorgensen, 1977). This process of flushing may have been more effective in the past during lower stands of sea level (higher fresh-water heads), as sea level fluctuated between present stand and 300 ft below present stand during the past 900,000 years (Meisler and others, 1985).

Under natural conditions, flow in unconfined and shallow-confined aquifers responds to slow changes in sea level relatively quickly. If fresh-water heads are high enough, all salt water in an aquifer open to the sea may be flushed out and a fresh-water spring may exist where this aquifer unit crops out at the sea floor. If fresh-water heads are lower than salt-water heads, a reversal in flow occurs, with salt water replacing fresh water. If fresh-water heads are not large enough to replace the salt water, a dynamic equilibrium will exist between the sea water and the fresh water (Hubbert, 1940). Kohout (1960) demonstrated that the dynamic equilibrium involves flowing sea water in the Biscayne aquifer of Florida, whereby sea water is returned to the sea. A transition zone, or zone of diffusion of variable thickness, exists between the two bodies of fresh water and salt water (Fig. 33). Tidal action and its impact on salt-water movement, as well as changes in the fresh-water potentiometric surface due to variation in fresh-water recharge or discharge, are the principal mechanisms governing the position and shape of this transition zone (Cooper and others, 1964).

Conditions may be somewhat different in deep confined aquifers, with regional ground-water flow reacting more slowly to changes in sea-level elevations. Such a case of slow equilibration, and the resulting present landward flow of saline water, was proposed by Meisler (1989) for deep aquifers in the Northern Atlantic Coastal Plain. Meisler and others (1985) estimated that the interface reflects a sea-water level between 50 and 100 ft below current level and that the interface adjusts to present sea level by

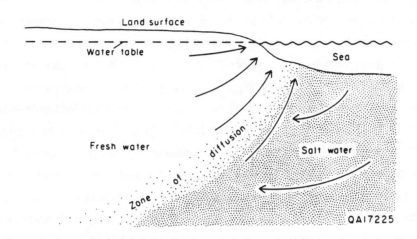

Figure 33. Typical ground-water flow patterns along coast lines, with circulation of salt water from the sea and back to the sea within a zone of diffusion (from Cooper and others, 1964).

moving inland at a rate of approximately one foot per year. Depths to top and the thickness of the interface or transition zone varies widely. For example, along the coast of the Northern Atlantic Plain, the transition zone of salt water containing between 250 mg/L and 18,000 mg/L is between 400 ft and 2,200 ft thick, reflecting long-term variations in the position of the fresh-water and salt-water zones (Meisler, 1989). This zone is thinnest in areas of fresh-water discharge (along coastal rivers and bays) because of low fresh-water heads in those areas. The top of the transition zone is shallowest in the southern part of the Northern Atlantic Plain, with the 250 mg/L contour at 200 to 400 ft below sea level. In the northern part of the Atlantic Plain, the same contour is at 2,000 ft below sea level, as fresh water extends farther onto the continental shelf in the north than in the south (Meisler, 1989).

The transition zone between fresh water (Cl <250 mg/L) and sea water (Cl >18,000 mg/L) in the Biscayne Aquifer at Miami, Florida, is only approximately 50 ft thick, reflecting the influence of tidal fluctuations and fresh-water discharge fluctuations as well as hydraulic connection between sea water and fresh water (Kohout, 1960). Within the transition zone some of the sea water may again be discharged to the sea (Fig. 33). Kohout (1960) suggested that under certain conditions approximately 20 percent of the intruding sea water discharges back to the sea through the transition zone in the Biscayne aquifer of Florida. Because of this flow in the transition zone, sea water must constantly intrude the aquifer (Cooper and others, 1964). Under certain conditions such as low fresh-water heads and inland flow of sea water along sinkholes and solution openings in carbonate aquifers, this mixing between fresh water and salt water may be encountered far enough inland to cause discharge of the brackish transition-zone water along the coast (Fig. 34) (Stringfield and LeGrand, 1969; Cotecchia and others, 1974).

Sea-water intrusion is not confined to the lower parts of aquifers, but also can occur in the upper, shallower sections of an aquifer, where strong storms cause flooding of coastal zones. Such a scenario is known along the coastal dunes aquifer of the northwestern United States, where winter storms create a several foot thick sea-water zone overriding fresh ground water (Magaritz and Luzier, 1985). The natural position of the interface between ground water and sea water is affected by tidal action, with fluctuations on a sandy beach vertically near 5 ft and horizontally 200 ft (Urish and Ozbilgin, 1985). This can create a mounding of sea water and consequently an effective mean sea level of close to 2 ft higher than local mean sea level (Urish and Ozbilgin, 1985).

The natural equilibrium of shallow coastal aquifers can be disturbed when the flow of fresh water decreases either in response to heavy pumping from water wells or in response to decreased recharge to the aquifer. Increased urbanization along United States coastlines have caused both of these mechanisms to occur in many places. Large areas have been eliminated as recharge points and today much water that used to recharge aquifers ends up in storm sewers. At the same time, pumpage of ground water from coastal aquifers in many areas has lowered water levels from originally above sea level and flowing conditions to sometimes tens or hundreds of feet below sea level, resulting in a reversal of ground-water flow and the potential of salt-water intrusion in the pumped area if heads are below the

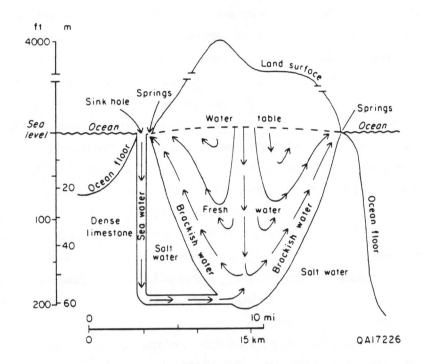

Figure 34. Schematic cross section showing mixing mechanism of sea water with fresh water through sinkholes and solution openings in a carbonate aquifer (from Stringfield and LeGrand, 1969).

85

submarine outcrop of the pumped unit or if the transition zone between fresh water and salt water falls within the cone of depression. For example, the city of Galveston used to produce fresh water from a 800-ft-deep well on Galveston Island but had to abandon the well in 1896 because of salinization. A new well field had to be established 15 miles inland (Turner and Foster, 1934). This early occurrence of sea-water intrusion may not be typical, but as ground-water withdrawals have increased in response to expanded demands, forced abandonments of water wells due to salinization have become widespread in many coastal areas. It is estimated that one to two public wells and 20 to 30 domestic wells are lost each year in Delaware as a result of salinization due to sea-water intrusion (Atkinson and others, 1986).

Increased domestic, agricultural, and industrial consumption of surface water from coastal streams can also lead to intrusion of sea water upstream. Where pumpage from river alluvium is high, subsequent intrusion of salt water into ground water can occur where the waterway is a losing stream. Dredging of waterways can enhance this process whenever a formally sealed-off permeable unit, such as permeable river alluvium, is encountered during dredging (Bruington, 1972). Such a case has been described by Barksdale (1940) for the Parlin area, New Jersey, where pumpage has decreased heads from originally flowing conditions to as much as 70 ft below land surface. Dredging of a new canal for shortening of the navigation route along the South River allowed salt water to enter through the canal bottom into the local aquifer, where it advanced at a rate of approximately one mile per six years (Barksdale, 1940; Schaefer, 1983). The sea-water intrusion problem can be aggravated by leaky or corroded well casings through which saline water can migrate to fresh-water aquifers (Miller and others, 1974; Monterey County Flood Control & Water Conservation District, 1989).

Another mechanism that adds salinity preferentially to coastal aquifers is sea spray during violent storms. This mechanism causes chloride concentrations in the tens of milligrams per liter in precipitation and deposits large amounts of sea salt along coastlines. For example, approximately 2.7 million tons of sea salt are deposited annually over the entire area of Japan (Tsunogai, 1975).

3.3.2. Chemistry of Sea Water

Sea water contains approximately 35,000 mg/L of dissolved solids, with chloride and sodium combining for 84 percent of the total concentration (Table 4). Salinity is somewhat higher in the Atlantic Ocean (36,900 mg/L at latitude 25°N) than in the Pacific Ocean (33,600 mg/L at latitude 40°N), depending on local circumstances such as continental influence, degree of evaporation, and oceanic currents (Custodio, 1987). Much larger differences can be found elsewhere. For example, inflow of fresh water reduces the salinity of the Baltic Sea to between 3,000 to 8,000 mg/L, whereas evaporation increases salinity in some areas of the Mediterranean and the Red Sea as high as 45,000 mg/L (Custodio, 1987). The influence of fresh-water inflow on water quality is illustrated by samples taken at increasing distance

Table 4. Concentration of major and significant minor chemical constituents in sea water (from Goldberg and others, 1971; Hem, 1985) (concentrations in mg/L).

Constituent	Concentration	Percent	Total Percent
Chloride, Cl	19,000	54	54
Sodium, Na	10,500	30	84
Sulfate, SO_4	2,700	8	92
Magnesium, Mg	1,350	4	96
Calcium, Ca	410	1	97
Potassium, K	390	1	98
Bicarbonate, HCO_3	142	0.4	98.4
Bromide, Br	67	0.2	98.6
Strontium, Sr	8	0.02	98.6
Silica, SiO_2	6.4	0.02	98.6
Boron, B	4.5	0.01	98.6
Fluoride, F	1.3	0.003	98.7
Iodide, I	0.06	0.00017	98.7

from the shore along the Texas Gulf Coast (Table 5). Chloride concentrations increase from 3,200 mg/L at the mouth of the Houston Ship Channel at Galveston Bay, to 13,000–14,000 mg/L three miles offshore and to 18,000 mg/L 25 miles offshore (Jorgensen, 1977).

As in most natural waters, the sum of all major cations and anions in sea water amounts to more than 98 percent of the total concentration of dissolved constituents (Table 4). However, the percentage of individual constituents is very different from that in fresh water and many other natural waters. In sea water, sodium is the dominant cation and chloride is the dominant anion, whereas in most fresh waters calcium is the dominant cation and bicarbonate or sulfate are the dominant anions. Even more striking is the difference in relation between calcium and magnesium concentrations. In sea water, the weight ratio of Ca/Mg is approximately 0.3. In contrast, in most fresh waters the Ca/Mg weight ratio is greater than 1.0. Also, potassium concentrations in most fresh waters are smaller than calcium concentrations by one order of magnitude or more, whereas they are nearly the same in sea water (Table 4).

Water samples from various sea-water intrusion sites suggest two mechanisms that alter the composition of the intruding sea water (Fig. 35). On bivariate plots of Ca versus Cl, K versus Cl, and Na versus Cl, cation exchange is suggested by some of the samples, as calcium content is greater and sodium and potassium contents are smaller than in the well-defined mixing trends indicated by other samples and by other constituent plots (Fig. 35). Mixing trends are less well defined at chloride concentrations of less than 1,000 mg/L because local fresh-water variations dominate over the relatively uniform sea-water composition at low concentrations.

3.3.3. Examples of Geochemical Studies of Sea-Water Intrusion

Whenever sea-water intrusion is the only source of salt water in a given area, identification of this salinization source in an affected well is indicated by an increase in total dissolved solids and possibly all major cations and anions. This recognition poses little difficulty. The most-used tracer of simple sea-water intrusion scenarios is the chloride ion, which is the most conservative, natural constituent in water once it is in solution. After establishing the base-line (background) chloride concentration in a given area from historical data or from wells that are not yet affected by salt water, periodic sampling of monitoring wells and analysis of chloride is being done along many areas along the United States coast where the potential of sea-water intrusion is feared. The importance of knowledge of background levels is illustrated in sometimes subtle chloride increases, such as those measured in samples from Middlesex County, New Jersey, where intrusion has been observed for the past 40 years. During the period of 1977 to 1981, chlorinity increased in some wells already affected by intrusion as well as in others that had shown background levels of less than 10 mg/L dissolved chloride in 1977 (Table 6) (Schaefer, 1983). Increases in chloride content from background levels of less than 5 mg/L to greater than 600 mg/L also indicated sea-water intrusion in response to heavy pumpage in some coastal wells of Monmouth County, New

Table 5. Landward changes in chemical composition of Gulf of Mexico water as a result of dilution (data from Jorgensen, 1977).

Source of Water	Date	SiO$_2$	Ca	Mg	Na	K	HCO$_3$	SO$_4$	Cl	TDS	Remarks
							Dissolved constituents in mg/L				
Houston Ship Channel, at Morgans Point	7/5/73	9.8	89	220	1,800	66	104	500	3,200	6,050	near surface
Gulf of Mexico, 3 miles offshore of Sabine Pass	10/9/74	0.5	280	840	7,100	330	140	1,800	13,000	23,400	at 1-ft depth
Gulf of Mexico, 3 miles offshore of Sabine Pass	10/9/74	0.4	300	880	7,900	250	140	2,000	14,000	25,400	at 24-ft depth
Gulf of Mexico, 25 miles offshore of Sabine Pass	10/9/74	0.1	370	950	9,500	330	146	2,500	17,000	30,700	at 1-ft depth
Gulf of Mexico, 25 miles offshore of Sabine Pass	10/9/74	0.2	380	940	9,800	320	148	2,400	18,000	31,900	at 42-ft depth

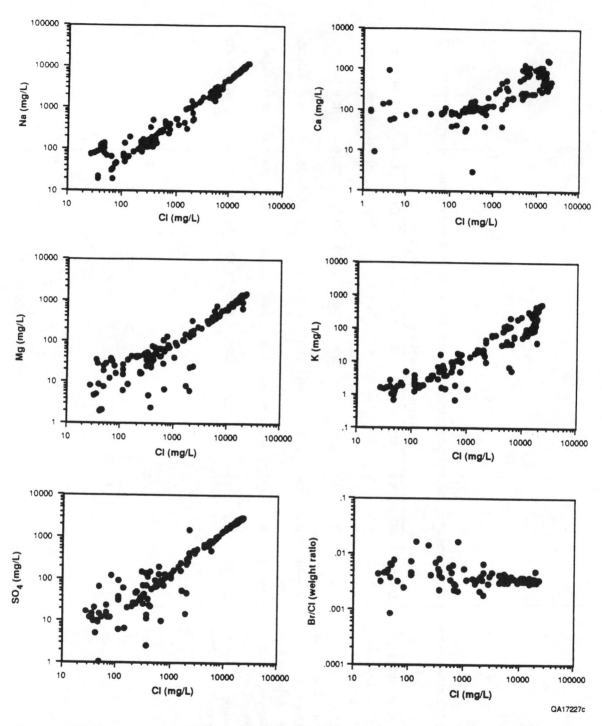

Figure 35. Bivariate plots of major ions and of Br/Cl ratios versus chloride for sea-water intrusion samples from California (data from Brennan, 1956), Israel (data from Arad and others, 1975), Hawaii (data from Mink, 1960), Spain (data from Price, 1988), and Texas (data from Jorgensen, 1977). Little scatter at chloride concentrations greater than approximately 1,000 mg/L indicates little variation between sea-water intrusion samples, with the exception of changes by cation-exchange reactions.

Table 6. Changes in chloride concentration between 1977 and 1981 in the Farrington aquifer of New Jersey as a result of sea-water intrusion (data from Schaefer, 1983).

Well name	Cloride concentration in mg/L	
	1977	1981
Perth Amboy WD 1A	6.3	32.0
Perth Amboy WD 2	49.0	54.0
South River BORO WD 2	7.2	12.0
South River Boro WD 5-77	7.0	12.0
South River Boro WD obs.	12.0	26.0
Sayreville Boro WD M	100.0	*190.0
Thomas and Chadwick 1	16.0	*42.0
El duPont - Parlin 1	7.5	45.0
El duPont - Parlin 3	47.0	96.0
Duhernal Water System 60F	680.0	1,300.0
NL Industries 3	7.6	*55.0

* Sampled in 1979

Jersey (Schaefer and Walker, 1981). There, intrusion was caused by heavy pumpage as hydraulic heads declined to 45 ft below sea level in the center of the cone of depression.

In the Southampton, New York, area, background chloride concentration of fresh water is less than 20 mg/L. Concentrations of 200 mg/L to 13,890 mg/L indicate mixing of fresh water with sea water (Anderson and Berkebile, 1976). A similar approximate limit, in some cases less than 20 mg/L in other cases less than 50 mg/L of chloride, above which sea-water intrusion is suspected, was suggested by Lusczynski and Swarzenski (1966) for Long Island, by Tremblay (1973) for Prince Edward Island, New York, by Wilson (1982) for the Floridan Aquifer of west-central Florida, by Kohout (1960) for the Biscayne aquifer of southeast Florida, and by Zack and Roberts (1988) for the Black Creek Aquifer of Horry and Georgetown Counties in South Carolina. On the other hand, rainwater near coastlines may contain chloride concentrations between 10 and 40 mg/L, and thus approach these background concentrations in some areas (Custodio, 1987). During hurricanes, ocean spray may affect coastal zones many miles inland, as documented during a 1938 storm event in New England, during which plants were damaged by spray as far as 45 miles inland from the coast (Hale, 1973). Background chloride concentrations may be higher in other areas, such as in the Monterey County area of California, where chloride levels range from 100 to 200 mg/L in uncontaminated wells (Fig. 36). Knowledge of background levels is useful in identifying the movement of an advancing intrusion front. For example, between 1977 and 1981 the continuous chloride increases in the Middlesex County area, New Jersey (Table 6) indicated a 0.2 to 0.4 mile inland migration of the transition zone between fresh water and salt water in just those four years (Schaefer, 1983).

In case sea-water intrusion is not the only potential source of salinity in a given area, differentiation of the respective salinization sources is more complicated and requires additional tracers or other tracers than chloride. In the Castroville area of Monterey County, California, two sand and gravel aquifers (180- and 400-ft aquifers) are affected by sea-water intrusion due to heavy pumpage (Monterey County Flood Control & Water Conservation District, 1989). Sea water has intruded the shallow unit up to five miles landward beginning in 1944 and the deep unit up to one mile landward beginning in 1959, causing chloride concentrations in water wells to increase from background values of 100–200 mg/L to greater than 1,000 mg/L (Fig. 36). Figure 37 illustrates this landward advancement in a plot of distance of the 500 mg/L chloride contour from the coastline over time for the two aquifer units. Leakage of salt water can occur from the shallow aquifer to the deep aquifer along wells penetrating both units. Using Piper diagrams, this local downward leakage of brackish water can be distinguished from regional sea-water intrusion (Monterey County Flood Control & Water Conservation District, 1989). The exchange of calcium and sodium between aquifer matrix and intruding sea water is reflected by a curve (Fig. 38) in the diamond-shaped field of the Piper diagram. This ion exchange proceeds in the intruding water until the aquifer sediments are saturated with sodium (Fig. 38 a1), after which no additional calcium is released from the matrix (Fig. 38 a2). Therefore, sea water that intrudes the aquifer behind this initial front will not exhibit the

92

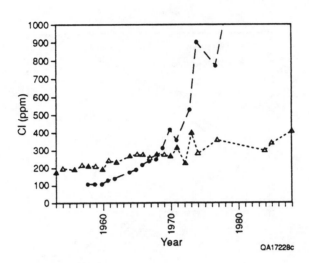

Figure 36. Increase in chloride concentration with time in coastal wells in Monterey County, California, suggesting sea-water intrusion (from Monterey County Flood Control & Water Conservation District, 1989).

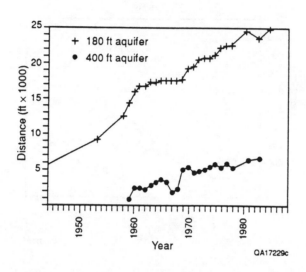

Figure 37. Landward movement of sea water mapped by the position of the 500 mg/L chloride contour in two shallow aquifers, Monterey County, California. Since 1944, sea water has advanced approximately 20,000 ft in the 180-ft aquifer to a position 25,000 ft inland from the coast (from Monterey County Flood Control & Water Conservation District, 1989).

94

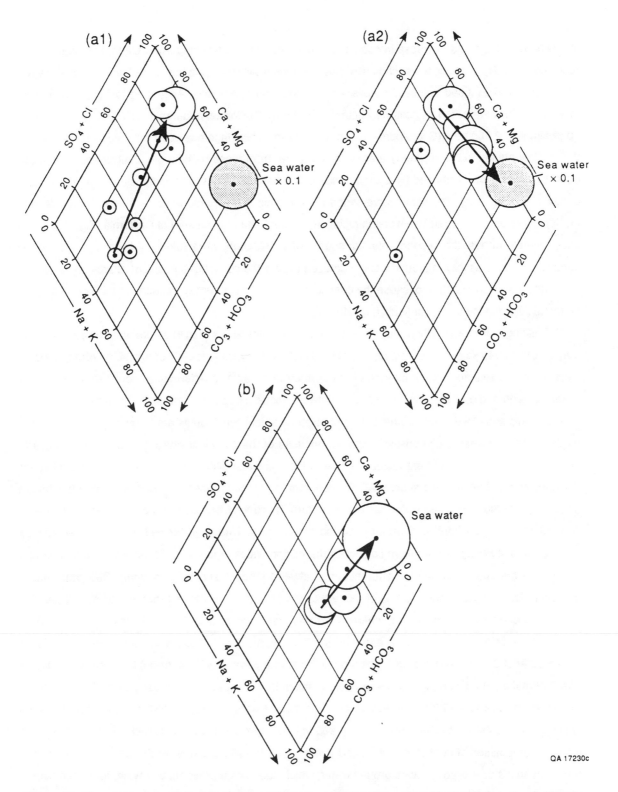

Figure 38. Piper diagram of chemical composition of coastal saline water, Monterey County, California. Regional sea-water intrusion is characterized by ion exchange (a1) and mixing (a2), whereas local mixing of intruded sea water with fresh water along boreholes is characterized by simple mixing (b) (modified from Monterey County Flood Control & Water Conservation District, 1989).

95

increase in calcium and decrease in sodium as observed in the advancing front. Often, this stage is not reflected in readily available water-chemistry data because wells producing this kind of water most likely will have ceased to be used unless the wells function as monitoring wells. According to Piper and others (1953), sodium-saturated conditions occur in Southern California aquifers when TDS levels reach approximately 2,000 ppm (Cl of approximately 1,000 ppm). This stage has been reached in some parts of the shallow aquifer in the Monterey area. Leakage water from these parts in the shallow aquifer into the deep aquifer are characterized by a simple mixing line between sea-water composition and fresh-water composition (Fig. 38b). The same curve-shaped relationship in the diamond field was shown by Brennan (1956) for waters affected by sea-water intrusion in the Manhattan Beach area of California (Fig. 39). In this case, oil-field brines, the second potential source of salinization, were ruled out in favor of sea-water intrusion because salinity in ground water increased toward the coast but not toward nearby oil fields. Oil-field brine was also ruled out because the chemical composition did not correspond to mixing lines in the diamond field and the triangular fields (Fig. 39).

Flushing of formation water may be less effective in some areas than in others or some coastal wells may be drilled into saline water that represents old instead of modern sea-water intrusion. With respect to water-resource management it may be of importance to be able to differentiate between modern sea-water intrusion and past sea-water intrusion or between modern sea-water intrusion and formation water.

Howard and Lloyd (1983) pointed out that major chemical constituents are frequently inconclusive in differentiating between these possible sources because of the chemical similarities of these sources to present-day sea water. Different residence time of the salt water within the aquifer may hold the key for distinguishing between these sources. Tritium content in recent continental water exceeds a few tritium units (TU) whereas deep sea water has almost a zero TU content because of the long residence time of sea water (Custodio, 1987). Therefore, mixing of continental fresh water with sea water should be reflected in a lowering of the tritium content in the mixing water. Age dating of the waters using carbon isotopes is another possibility for differentiating between modern and old sea water. Sea water has a carbon-14 isotopic value close to that of modern organic matter (>80 percent modern; $\delta^{14}C = 0$ permille for recent, pre-bomb ocean water), whereas old sea water, connate or intruded, has a low carbon-14 content (Custodio, 1987). Hanshaw and others (1965) reported a carbon-14 value of +285 permille for ocean water at Jekyll Island, Georgia, and a range from +4.2 permille (±7.6 permille) for surface samples to −54.3 permille (±11.5 permille) for deep water masses from various parts of the North Atlantic Ocean. These values are very different to values of fresh ground water (−965 to −987 permille) and for deep saline ground water in the Claiborne Group (−968 to −981 permille) in the Brunswick, Georgia, area. The isotopic composition of contaminated ground water (−970 to −980 permille) in the area was identical to that of fresh water and deep saline ground water, which suggests that recent sea water is not the source of salinization in the contaminated water (Hanshaw and others, 1965).

96

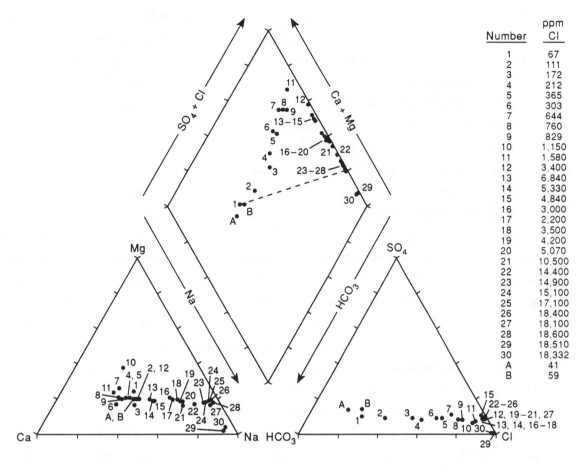

Number	ppm Cl
1	67
2	111
3	172
4	212
5	365
6	303
7	644
8	760
9	829
10	1,150
11	1,580
12	3,400
13	6,840
14	5,330
15	4,840
16	3,000
17	2,200
18	3,500
19	4,200
20	5,070
21	10,500
22	14,400
23	14,900
24	15,100
25	17,100
26	18,400
27	18,100
28	18,600
29	18,510
30	18,332
A	41
B	59

QA 17263c

Figure 39. Piper diagram of chemical composition of selected saline waters from the Manhattan Beach area, California (from Brennan, 1956). Cation exchange is suggested by the curve in the center field, with fresh water (No. 1) and ocean water (No. 28) as end points. Linear trends in the triangles also have ocean water as the high-Cl end member, as opposed to oil-field samples (No. 29, No. 30).

97

Age dating in combination with I/Cl ratios allowed Howard and Lloyd (1983) to differentiate three bodies of saline water as three separate intrusions into the Chalk limestone of eastern England. Modern radiocarbon ages were associated with minor iodide enrichment (I/Cl = 4×10^{-6}, as compared to I/Cl = 3×10^{-6} for sea water) and were interpreted as modern and active saline intrusion. Moderate iodide enrichment (I/Cl = 3×10^{-5}) was typical for shallow saline water related to Middle/Late Flandrian marine transgression and radiocarbon dates of 5,000 to 8,000 years. Deep saline water, dated at 7,000 to more than 21,000 years old and related to the Ipswichian interglacial stage, were characterized by high iodide enrichment (I/Cl = 10^{-4}). In this setting, the presence of iodide is an indicator for ground-water residence time as more iodide is leached out of the sediments over time (Fig. 40) (Lloyd and others, 1982). The amount leached from the sediments also depends on the sediment type, as iodide concentrations in sandstones are typically much lower (0.1–1 ppm) than those in argillaceous shales and limestones (>> 1 ppm); highest iodide concentrations are often associated with organic-rich marine deposits as well as with evaporites and caliche deposits (Lloyd and others, 1982). The ratio of I/Cl in sea water is approximately 500 to 1,000 times less than the value typically measured in rain samples from the Hawaiian atmosphere (Duce and others, 1965) and also one order of magnitude less than typical oil-field/deep-basin brine ratios and typical fresh water ratios (I/Cl >10^{-5}; Whittemore and Pollock, 1979). This difference may be useful to differentiate a sea water source from an oil-field/deep-basin brine source. The I/Cl ratio in sea-water is similar to the ratio found in many halite-dissolution brines (I/Cl <10^{-5}; Whittemore and Pollock, 1979; Richter and Kreitler, 1986a,b), which suggests that this ratio is not a good tracer for differentiating between these two sources. However, in combination with boron and barium, iodide was a useful tracer to distinguish sea-water intrusion from connate brine in a contamination study conducted by Piper and others (1953) in California.

Modern and old sea water or formation water may also differ in their stable isotope composition of oxygen-18 and deuterium as long as different climatic conditions existed during intrusion of the water (Custodio, 1987). By convention, sea water has a composition close to the zero permille value because "standard mean ocean water" (SMOW) is taken as a reference value against which all other samples are measured. Fresh ground water is typically isotopically lighter because of fractionation processes that occur during evaporation of sea water and condensation leading to precipitation. Water that has undergone evaporation before entering local ground water, however, may be heavier than sea water. Also, rock-water interactions at elevated temperatures tend to isotopically enrich the oxygen-18 content of the water (Custodio, 1987). These differences may allow differentiation of salinization sources.

In a similar fashion, the value of the stable carbon isotope ^{13}C differs between sea water and continental water. Mook (1970) reported $\delta^{13}C$ concentrations between 1.5 and 2.0 permille for sea water, whereas continental waters are isotopically lighter with ^{13}C of about –13 permille. According to B. B. Hanshaw (personal communication, 1976; in Jorgensen, 1977), a $\delta^{13}C$ value of –1 permille would be expected in sea water and a $\delta^{13}C$ range of –12 to –25 permille would be expected for ground water in the

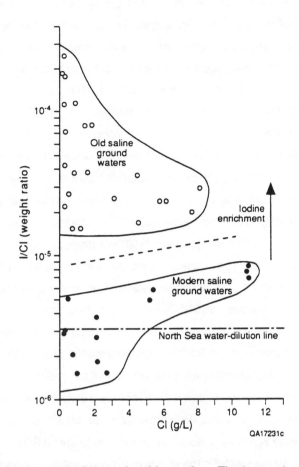

Figure 40. Iodide concentrations as indicator of residence time. The longer the residence time, the more leaching of iodide from sediments can occur, giving rise to higher concentrations than in waters with small residence times (modified from Lloyd and others, 1982).

Houston area. Testhole samples, ranging from −9.4 to −12.9 permille, appear to be inconclusive concerning a sea-water source (Jorgensen, 1977). Age dating, using tritium, and the use of the stable isotope ratios $\delta^{18}O$, δ^2H, $\delta^{34}S$, and $\delta^{13}C$ were also inconclusive with respect to the most likely source of salinization of ground water along the Houston Ship Channel. Instead, major cations and anions, depicted in Piper diagrams and Stiff diagrams, appeared to best reflect the potential sources of salinity in that area (Houston Ship channel in the shallow subsurface, formation water in deeper units). Although water samples had been analyzed for many minor and trace elements such as Br, I, Li, Fe, B, K, F, and SiO_2, Jorgensen (1977) did not make use of or discuss these constituents as possible tracers.

Another ratio indicative of residence time may be the ratio of Ca/Mg. The Ca/Mg weight ratio of 0.3 in sea water is very small when compared to the characteristic value of >1 found in most fresh waters and brines. Ion exchange between intruding sea water and fresh-water aquifer matrix results in an increase in the Ca/Mg ratio when compared to the ratio of sea water. This increase may be more pronounced in older waters than in modern waters, which led Sidenvall (1981) to propose that saline water in the Uppsala, Sweden, area is fossil ground water representing the last sea transgression. The ratio of Cl/SO_4 was used by Snow and others (1990) to distinguish sea-water intrusion from road-salt contamination in Maine, based on the fact that sulfate concentrations and Cl/SO_4 ratios are substantially lower in road-salt affected well waters than in sea-water affected well waters. This ratio can also be used to distinguish modern sea-water intrusion from previous intrusion in waters having chloride concentrations greater than 500 mg/L, as done by Pomper (1981) in the Netherlands. As a result of sulfate reduction, older saline waters are characterized by Cl/SO_4 ratios that are higher than those found in modern seawater. In a similar fashion, Martin (1982) explained the source of saline water in the Santa Barbara, California area as ocean water that had undergone ion exchange and sulfate reduction. In addition to the Cl/SO_4 ratio, the ratios of B/Cl and Ba/Cl appeared to indicate that sea water is the source of the deep saline ground-water in that area (Martin, 1982). The SO_4/Cl ratio also seems to indicate mixing between sea water and recharge water in the Floridan aquifer of southwest Florida, as shown by samples that plot close to the theoretical mixing line in a plot of SO_4/Cl ratios versus SO_4 concentrations (Fig. 41) (Steinkampf, 1982). However, gypsum-anhydrite solution causes sulfate concentration to increase downgradient in some of the samples, limiting the use of the SO_4/Cl ratio as a tracer of mixing. Because of the generally high degree of water–rock interaction in this carbonate-dominated aquifer which governs concentrations of calcium, magnesium, and bicarbonate and masks the source of salinity, the conservative nature of the chloride and bromide ions was used by Steinkampf (1982) to demonstrate that dilution of marine-like ground water is a significant mechanism in the evolution of saline ground water in this aquifer. Past marine inundations, after saline interstitial waters had been flushed from the sediments, are the probable source of salinity (Steinkampf, 1982).

The Br/Cl weight ratio in sea water is approximately 3.3×10^{-3}, which is comparable to most fresh waters. Therefore, mixing of sea water and fresh water may cause only a slight change in Br/Cl ratio in the

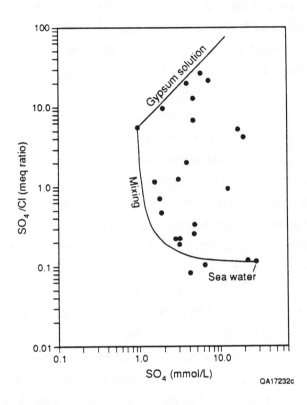

Figure 41. Bivariate plot of SO_4/Cl ratios and SO_4 concentrations for coastal saline ground water of southwest Florida. Sea-water intrusion is indicated in those samples that plot close to the theoretical mixing line between sea water and fresh water (modified from Steinkampf, 1982).

mixing water when compared to the fresh-water value. For example, Brennan (1956) reported Br/Cl ratios of 3.4×10^{-3} for Pacific Ocean water and a range of 3.2×10^{-3} to 3.7×10^{-3} for sea-water intrusion samples in the Manhattan Beach area, California. However, because of the conservative nature of Br and Cl, this ratio can often be used to differentiate between salt-water sources. For example, halite-dissolution brine is typically characterized by Br/Cl ratios smaller than 5×10^{-4}, which is one order of magnitude smaller than the value in sea water. On the other hand, oil-field brines often have ratios of Br/Cl that are significantly higher than sea-water and halite-dissolution brine values (Whittemore and Pollock, 1979; Richter and Kreitler, 1986a,b; Whittemore, 1988). Therefore, the Br/Cl ratio may be useful to differentiate between sea-water intrusion and oil-field contamination in coastal settings that include oil fields. In settings that include storage and application of salt for road deicing, such as along the coast in the northeastern part of the United States, the Br/Cl ratio can also be used to distinguish sea-water intrusion from road-salt contamination. Saline ground water originating from past or current intrusion of sea water contains detectable bromide with Br/Cl ratios similar to sea water, whereas saline ground water originating from road salt typically contains little bromide with Br/Cl weight ratios less than 1×10^{-3} (Snow and others, 1990).

Ion exchange and carbonate dissolution are major chemical reactions that alter the composition of intruding sea water. This is the case even in aquifers that contain only small amounts of clay and carbonates, as shown in a study of rock-water interactions and seawater-freshwater mixing in coastal dune aquifers of Oregon. Magaritz and Luzier (1985) found that of all the major ions, chloride was the only one to act conservatively although the aquifer consisted of rather uniform quartz sand with only minor amounts of silt, clay, and some organic matter. Sulfate reduction, oxidation of plant material, formation of authigenic K-feldspar, and Ca-Na base exchange on clay minerals cause the composition of the saline water to differ considerably from the theoretical composition of diluted sea water in the mixing zone. Nadler and others (1981) summarized processes occurring in the transition zone within sandy sediments as being (a) dilution of sea water, (b) Ca-Mg exchange, (c) Na-Ca or Na-Mg base exchange, and (d) sulfate reduction. The same was shown by Mink (1960), who studied concentration changes in sea water as the water moves through calcareous and alluvial deposits on the ocean bottom before entering the basaltic aquifer of southern Oahu, Hawaii. Even at a very small degree of dilution, indicated by a chloride concentration of 17,600 ppm in the intruded water as compared to 19,000 ppm in open sea water, changes in major cations are very noticeable, with the calcium concentration increasing by more than 100 percent and the potassium concentration decreasing by more than 50 percent (Mink, 1960). Similarly, sea-water intrusion samples in the coastal zone of Fuji City, Japan, do not reflect the theoretical mixing composition of sea water and fresh water. Instead, ion exchange and solution-precipitation reactions cause (1) higher Ca content, (2) slightly higher SO_4 content, and (3) lower Na, K, and HCO_3 content than expected from simple mixing (Ikeda, 1967). The increase in Ca content with time was used by Jacks (1973) to distinguish old sea water from young sea water, based on the assumption that old water had more time for ion exchange reactions to occur and, therefore, would be characterized by higher Ca contents than younger sea water. Under certain circumstances, manganese concentrations may be indicative of ion exchange. This is the case in sea-water intrusion samples from Manhattan Beach, California, in which sodium ions of the intruding sea water are exchanged for manganese in the sand and gravel aquifer. Largest manganese concentrations appear to be associated with Mg/Ca ratios close to or slightly greater than one (Brennan, 1956).

Dilution diagrams were used by Howard and Lloyd (1983), who investigated the relationship of three groups of saline ground water in east-central England (Fig. 42). On these plots, data points close to the theoretical mixing line between sea water and local ground water indicate simple mixing, whereas data points away from mixing lines indicate chemical reactions, such as mineral dissolution and precipitation, ion exchange, and sulphate reduction. It is interesting to note that Howard and Lloyd (1983) did not at first attempt differentiating saline waters by the use of major chemical constituents but instead used isotopes and the I/Cl ratio to distinguish three groups. Only because of this grouping did dilution diagrams allow an explanation of the hydrochemical evolution of the individual groups that otherwise would have only resulted in a big cluster of data points.

3.3.4. Reaction Characteristics of Sea-Water Intrusion

The chemical composition of sea water changes as it intrudes a fresh-water aquifer. Changes occur in response to mixing and chemical reactions as summarized in Figures 43 and 44 and below. These changes are most pronounced within the initial sea-water front that mixes with fresh water. Subsequent intrusion deviates little from sea-water composition.

Mixing: Mixing of fresh water and sea water occurs within a transition zone, which is characterized by chloride concentrations from just above background concentration values to just less than sea-water concentration. The front of this transition zone is characterized by ion exchange as discussed below. Behind the ion-exchange front, simple dilution characterizes deviation of the brackish water from sea-water composition. This can easily be identified on trilinear plots in the straight-line relationship between data points. On bivariate plots of major cations and anions versus chloride, data points plot close to the theoretical mixing line between local fresh water and sea water.

Ion Exchange: Clay minerals, especially montmorillonite, have free negative surface charges that are occupied by cations in direct proportion to the abundance of cations in the water and to the sorption characteristics of the cations and the minerals. In a typical fresh-water aquifer, these sites are saturated mainly with calcium ions, whereas in a typical salt-water aquifer, the sites are occupied mainly by sodium ions. Whenever the relation of calcium to sodium in the water changes, for example in response to sea-water intrusion into a fresh-water aquifer, ion exchange will occur whereby sodium will be taken out of solution and calcium will be released from mineral exchange sites. Magnesium and potassium may also be exchanged for calcium but the Na-Ca exchange is the most significant one. For example, Howard and Lloyd (1983) attributed more than 96 percent of the base exchange in the Chalk aquifer of east-central

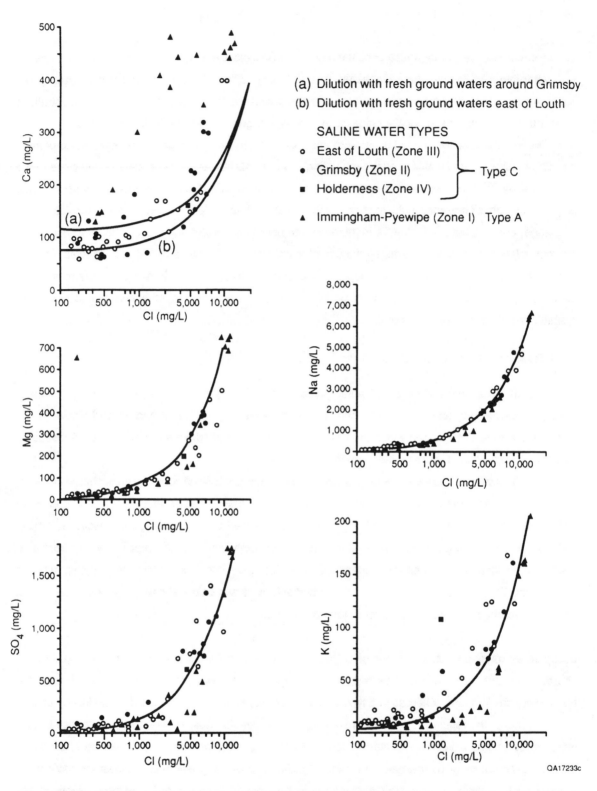

(a) Dilution with fresh ground waters around Grimsby

(b) Dilution with fresh ground waters east of Louth

SALINE WATER TYPES

○ East of Louth (Zone III)

● Grimsby (Zone II)   Type C

■ Holderness (Zone IV)

▲ Immingham-Pyewipe (Zone I)   Type A

QA17233c

Figure 42. Dilution diagrams of major ions with theoretical mixing lines between sea water and local fresh water. Simple mixing is indicated for samples plotting close to theoretical mixing lines. Deviations from these lines suggest additional chemical reactions (from Howard and Lloyd, 1983).

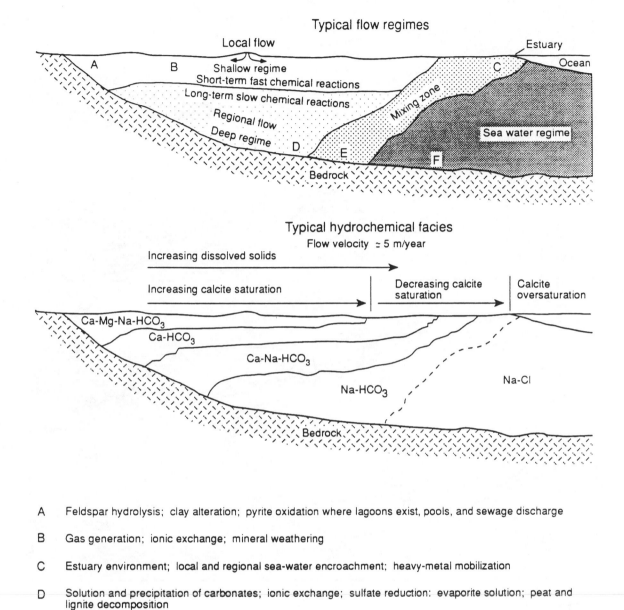

Typical flow regimes

A   Feldspar hydrolysis; clay alteration; pyrite oxidation where lagoons exist, pools, and sewage discharge

B   Gas generation; ionic exchange; mineral weathering

C   Estuary environment; local and regional sea-water encroachment; heavy-metal mobilization

D   Solution and precipitation of carbonates; ionic exchange; sulfate reduction: evaporite solution; peat and lignite decomposition

E   Mineral alterations; cement deposition; dolomitization

F   Man-made effects where deep injection of waste water occurs

QA17235c

Figure 43. Relationship between flow regimes and hydrochemical facies in ground water in a big coastal plain (from Back, 1966, and Custodio, 1987).

105

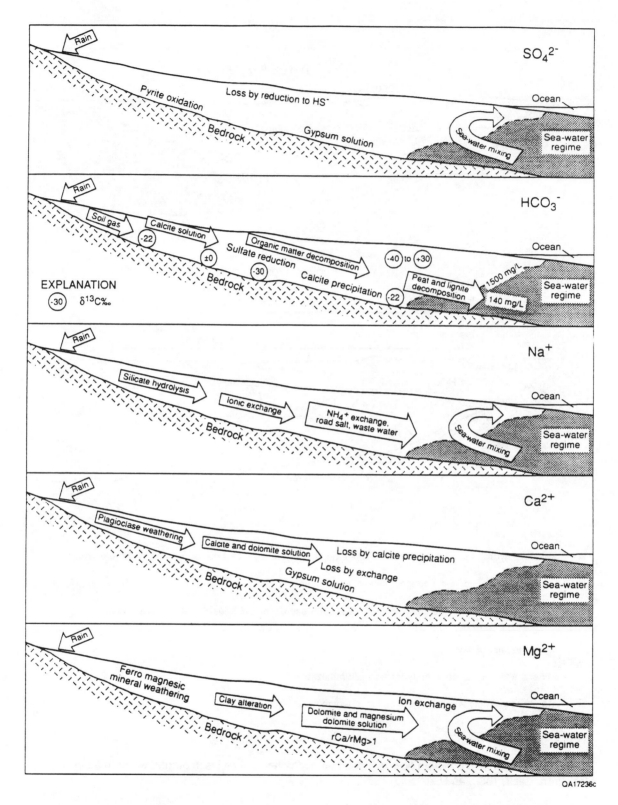

Figure 44. Geochemical processes and changes in ionic composition along the flow path in a coastal aquifer (from Custodio, 1987).

106

England to Na-Ca exchange. This exchange is assumed to be instantaneous (Kafri and Arad, 1979). Intrusion of fresh water into a salt-water aquifer will cause the opposite ion exchange. Table 7 summarizes major changes on ionic ratios as a result of these changes in water facies. The chloride concentration is not affected by ion exchange, which makes the Na/Cl ratio a potential tracer of intrusion. If sea water intrudes a fresh-water aquifer, Na/Cl ratios will decrease from ratios often greater than one to ratios often less than the value in sea water. In contrast, if fresh water replaces marine water or washes out marine sediments, very high Na/Cl ratios can result (Custodio, 1987). Figure 39 illustrates this process of ion exchange in the early part of the intruding sea-water front as plotted on a Piper diagram. Ion exchange between calcium and sodium characterizes the cluster in the cation triangle (samples 1 through 12, representing the lowest chloride values), whereas mixing characterizes the straight line in the anion triangle as chloride content increases. The diamond-shaped field reflects the slight increase in Ca+Mg, the matching decrease in Na+K, and the high increase in Cl+SO_4 percentages (Fig. 39).

No changes in the Na/Cl ratio will occur in water that intrudes behind the front of ion exchange because all the exchange sites are already occupied. Therefore, the Na/Cl ratio should approach the ratio of sea water (0.85 molar ratio), which differs from the typical ratio of halite-dissolution brines (0.64 molar ratio) and from the small ratio characteristic for many oil-field/deep-basin brines (<0.50 molar ratio). The degree of change that occurred because of ion exchange may not only indicate the position within the intruding front (Table 7), but also the timing of the intrusion. Recent sea-water intrusion would be expected to be associated with data points predominantly showing ion exchange, whereas old sea-water intrusion would be expected to include many data points with little or no evidence of ion exchange (Howard and Lloyd, 1983).

Carbonate Dissolution: Mixing of fresh water and sea water, both saturated with respect to calcium carbonate, can result in a mixing water that is undersaturated with respect to calcium carbonate (Hanshaw and others, 1971; Back and others, 1979; Custodio, 1987). This mixing water can dissolve carbonates and thus calcium and bicarbonate concentrations will increase. Additional calcium carbonate dissolution may occur in the presence of sulfate reduction of organic-rich sediments because of the associated change in the pH and CO_2 contents of the water.

Sulfate Reduction: Sea water is relatively high in dissolved sulfate content (Table 4). Under reducing conditions in ground-water systems, and with the presence of compounds that can be oxidized (that is, organic matter) as well as of reaction catalysts (for example, bacteria or isolated enzymes), sulfate will be reduced according to the equation $CH_2O + 1/2 SO_4^{2-} \longrightarrow HCO_3^- + 1/2HS + 1/2H^+$ (with $HS^- + H^+ \longrightarrow H_2S$ at pH <7) Freeze and Cherry, 1979). This results in a decrease in sulfate concentration relative to the sea-water composition.

Bromide: Sea-water intrusion may lead to bromide concentrations greater than a few milligrams per liter in coastal aquifers. This has a big potential impact on future drinking-water resources because elevated bromide concentrations make water treatment more difficult. Traditional chloride treatment or

Table 7. Changes in ionic ratios due to ion exchange relative to the position of the intruding water.

	Recent Salt-water Intrusion		Fresh-water Intrusion
	(a) Advancing front	(b) Behind advancing front	
Ca/Na	increase	no change	decrease
$\Delta\lvert Ca+Mg\rvert/\Delta\lvert Na\rvert$	constant	no change	constant
Na/Cl	decrease	no change	increase

ozonation of high-bromine water causes the formation of bromoform and other brominated trihalomethanes, which may, at bromide concentrations greater than 2 mg/L in the water, exceed allowable concentration in future (1992) standards (McGuire and others, 1989, U.S. Water News, 1990a).

Minor Constituents: As sea water passes through muds on estuary or ocean bottoms, it may become enriched in those minor constituents that are typically concentrated in those muds, such as iodide, strontium, and fluoride (Lloyd and Heathcote, 1985).

3.3.5 State-by-State Summary of Sea-Water Intrusion

Sea-water intrusion is not a new phenomenon and has been reported to occur in every coastal state of the United States. For example, the city of Galveston had to abandon one of their water-supply wells as early as in 1896 (Turner and Foster, 1934) and in Florida, intrusion has been experienced along most of the Gulf and Atlantic Coast (Atkinson and others, 1986). The following section lists the published occurrences for sea-water intrusion for all the states with marine shorelines. This discussion includes sea-water intrusion due to anthropogenic activities but also mentions the occurrence of saline ground water in coastal aquifers related to natural processes, such as inclusion since deposition or past intrusions during times of higher sea-water levels. This compilation is a result of literature review.

Alabama: Sea-water intruded fresh-water aquifers from Fort Morgan to Gulf Shores (Atkinson and others, 1986), and overdevelopment of a well field in the Mobile-Gulf Coast region caused lateral intrusion of saline water from the Mobile River into Pleistocene sand-gravel aquifers (Miller and others, 1977). Sea-water contamination of wells in Baldwin County has been caused by overproduction of a fresh-water zone, sea-water flooding and natural sea spray, and leakage from salt-water ponds (Chandler and others, 1985).

California: In southern California, the potentiometric surfaces of coastal aquifers were above sea level at the turn of the century but gradually decreased in the following decades. Because of heavy developments in the 1920's and 1940's water levels have declined in some areas to as much as 70 ft below sea level (Brennan, 1956). In the Manhattan Beach area of Los Angeles, most of the original water wells were abandoned after 1940 when salinity of ground water increased due to sea-water intrusion (Poland and others, 1959). Banks and Richter (1953) summarized: known sea-water intrusion areas (A), threatened sea-water intrusion areas (B), and potential sea-water intrusion areas (C) in California as being: (A) Mission, San Luis Rey, Santa Margarita, Coastal Plain Orange County, Coastal Plain Los Angeles County, Malibu Creek, Trancas Creek, Morre Bay, Salinas, Coastal strip Salinas Valley and Pajaro Valley, Pajaro, Santa Clara, and Sacramento–San Joaquin Valley between Pittsburgh, and Antioch; (B) Tia Juana, Otay, Santa Clara River, Carpenteria, Goleta, Arroyo Grande; and (C) Sweetwater, San Dieguito, San Onotre, San Mateo, San Juan, Ventura River, Santa Ynez River, Santa Maria River, Carmel, Ygnacio Clayton, Napa-Sonoma, Petaluma, Eel River, Mad River, and Smith River. Of those, the most serious sea-water intrusions have occurred in the West Coast Basin of Los Angeles County, East Coastal Plain

Pressure Area of Orange County, Petaluma Valley in Sonoma County, Napa-Sonoma Valley, Santa Clara Valley in the San Francisco Bay area, Pajaro Valley in Monterey and Santa Cruz Counties, Salinas Valley in Monterey County, Oxnard Plain Basin in Ventura County, and Mission Basin in San Diego County (Fuhriman and Barton, 1971). In 1975, the 14 known (A) and 14 suspected (B) areas of sea-water intrusion were (A) Eel River Valley, Petaluma Valley, Napa-Sonoma Valley, Santa Clara Valley, Pajaro Valley, Elkhorn slough area, Salinas Valley pressure area, Morro Basin, Chorro Basin, Los Osos Basin, Oxnard Plain Basin, West Coast Basin (Los Angeles County), San Luis Rey Valley, and Mission Basin, and (B) Russian River Basin, Drakes Estero Basin, Bolinas Lagoon Basin, San Rafael Basin, Suisum-Fairfield Valley, Sacramento-San Joaquin Delta, Tonitas Creek Basin, and San Diego River-Mission Valley Basin (Smith, 1989). Farrar and Bertoldi (1988) identified areas of salt-water intrusion around San Francisco Bay and the delta area, around Monterey Bay, Morro Bay and at the mouths of Petaluma, Sonoma, and Napa Valleys, where the salt water infiltrated from tidal channels. Diversion of stream flow from the Sacramento and San Joaquin Rivers has caused inland migration of salt water in the areas around Fairfield and Pittsburgh, whereas fresh-water pumping has caused sea-water intrusion and land subsidence in Alameda County. Pumping of water for irrigation purposes has led to sea-water intrusion near the mouths of the Pajaro and Salinas Rivers (Farrar and Bertoldi, 1988).

Connecticut: Lateral sea-water intrusion from harbors and tidal river estuaries due to heavy pumpage has contaminated many industrial and municipal wells in the Long Island Sound coastal area including the cities of New Haven and Bridgeport (Miller and others, 1974).

Delaware: Salt-water intrusion has been reported all along the Delaware coastline and is responsible for the abandonment of one to two public wells and 20 to 30 domestic wells each year (Atkinson and others, 1986). During dry periods, sea water has migrated up the Delaware River nearly as far as Philadelphia (U.S. Environmental Protection Agency, 1973). Lateral and vertical intrusion of salt water from the Delaware River and Delaware Bay and from the Atlantic has contaminated many well fields due to heavy pumpage from shallow aquifers and the dredging of impermeable soils (Miller and others, 1974). High chloride concentrations between 6,000 mg/L and 17,000 mg/L were measured in core samples obtained from wells along the Atlantic Coast of southeastern Delaware. The depths of wells from which core samples were obtained ranged from 5 to 60 ft below land surface (Woodruff, 1969).

Florida: The Florida problem of sea-water intrusion stems from a combination of permeable limestone aquifers, a lengthy coastline, and heavy pumpage in coastal areas. Intrusion was reported (in 1953) at 28 specific locations and some 18 municipal water supplies have been adversely affected since 1924. Interior drainage canals which lowered the water table and permitted sea water to advance inland by tidal action contributes to the problem (Bruington, 1972; U.S. Environmental Protection Agency, 1973). Sea-water intrusion is a permanent threat to the Biscayne Aquifer because the aquifer is unconfined, is hydraulically connected to the sea, is heavily pumped, and is extensively cut by a network of canals. Heavy pumpage has resulted in water levels below sea level near some well fields (Johnston and Miller, 1988).

Large cones of depression and sea-water intrusion due to heavy pumpage have been reported in many areas, such as Jacksonville, Tampa, and Miami. Intrusion on a smaller scale has been experienced along most of the Atlantic and Gulf Coasts (Atkinson and others, 1986). Miller and others (1977) listed some of the affected counties, including Escambia County (heavy pumping near Bayou Chico caused intrusion into sand and gravel aquifer and reversed flow gradient and induced salt-water migration from the Escambia River), Bay County (saline surface water leaked downward from bays contaminating two wells), Martin County (encroachment of water into shallow aquifer from St. Lucie River due to overpumpage of city wells), Pasco County (sea-water intrusion due to overpumpage of coastal wells), Charlotte County (sea-water intrusion from Gulf and estuaries and intrusion due to poor well construction and improper abandonment), Palm Beach County (inland flow of saline water through canals and intracoastal waterways during dry period caused contamination of Biscayne aquifer), Broward County (salt-water migration into Biscayne aquifer in coastal areas due to heavy pumpage and construction of canals), Dade County (dry period caused inland flow of salt water up the Miami canal and contaminated four water wells), and Miami (contamination of wells due to inland flow of salt water during dry periods and inflow of saline ground water due to heavy pumpage).

In 1943 in the Miami area, sea-water intrusion was restricted to wells within two miles of the coast except for some areas along tidal canals where intrusion was observed somewhat farther inland (Love, 1944). In Citrus County sodium and chloride concentrations exceed acceptable levels twofold and threefold, respectively, as a result of high pumpage of fresh water and seepage of salt water from the Gulf (U.S. Water News, 1990b). In the Floridan aquifer of west-central Florida, the transition zone between fresh water and saline water may be as far as 50 miles inland from the coast. Future ground-water withdrawal could lead to further landward movement of the salt-water front in that area at an average rate of approximately 0.35 ft per day (Wilson, 1982). In Pascola County, the interface between fresh and saline ground water, at 100 ft below sea level, is located approximately one to two miles inland (Reichenbaugh, 1972). Three encroachment mechanisms are responsible for that position: (1) lateral inflow through permeable limestone where the aquifer is in contact with sea water offshore, (2) leakage from tidal streams and canals in which sea-water intruded, and (3) upward movement of salt water in response to lowered hydraulic heads caused by pumpage of fresh water (Reichenbaugh, 1972).

Sea-water intrusion into fresh-water reaches of the Caloosahatchee River occurs between La Belle and Olga during low-flow periods as repeated injections through the lock chamber at W. P. Franklin Dam. In 1968, the upstream limits of water containing less than 250 mg/L of chloride were 11.4 miles from the dam in the deeper parts of the river and approximately five miles from the dam at shallow depth (Boggess, 1970).

Georgia: Sea-water intrusion has been reported for Chatham County (Savannah) and Glynn County (Atkinson and others, 1986).

Louisiana: Sea-water intrusion has occurred all along the coastal shores of Louisiana. In addition, the cities of Lake Charles and New Orleans have experienced severe cases of salt-water intrusion due to high pumpage of ground water (Atkinson and others, 1986). During periods of low flow, tidal water from the Gulf of Mexico invades coastal streams and fresh-water aquifers, such as the one in the Vermillion River area (Newport, 1977).

Maine: Overpumping has resulted in intrusion of tidal estuary waters into local aquifers south of Augusta, in Kennebec and Sagadahoc Counties. Several domestic water wells have been affected by salt-water intrusion, the problem fluctuating at times over the year (Atkinson and others, 1986). A 300-ft-deep well producing from the bedrock aquifer near the town of Bowdoinham, Sagadahoc County, was contaminated by salt water from the tidal reach of the Kennebec River, resulting in abandonment of the well (Miller and others, 1974).

Maryland: Sea-water intrusion has been reported in St. Mary's, Anne Arundal, Harfork, Dorchester, and Somerset Counties as well as on Kent Island. Overpumping, leaky well casings, and dredging activities appear to be responsible for ground-water contamination. Salty water from the Patapsco River estuary has intruded shallow fresh-water aquifers in the harbor district of Baltimore. Heavy pumpage and leaky casings also induced the inflow of salt water from Chesapeake Bay into wells at Joppatowne, Harford County, and Westover, Somerset County. Lateral and vertical intrusion of salt water from tidal river estuaries, enhanced by casing leaks in abandoned wells, has been reported in the areas of Cambridge, Dorchester County, Annapolis, Anne Arundal County, and the Solomons-Patuxent River, St. Mary's County. The inland limit of saline ground water in coastal plain aquifers are along the coast for the Cretaceous aquifer, along a line from Dorchester to the southwestern state-line corner with Delaware in the Tertiary aquifer, and along a line parallel to the Chesapeake shoreline in the south and approximately half way between the shore and the Maryland-Delaware state line in the north, crossing the state line just north of 39 degrees latitude (Miller and others, 1974).

Massachusetts: Sea-water intrusion has been reported in several areas in Massachusetts, including Bristol, Plymouth, and Barnstable Counties (Atkinson and others, 1986) as well as Provincetown, Scituate, and Somerset Counties (Newport, 1977).

Mississippi: Heavy pumpage and the resulting decline in water levels induced lateral intrusion of saline water from the Gulf and the Pascagoula River estuary at Moss Point into Miocene and the Citronelle Formations (Miller and others, 1977). Sea-water intrusion has also been reported in Hancock and Jackson Counties (Atkinson and others, 1986).

New Hampshire: Tidal waters have intruded aquifers in the Portsmouth area (Newport, 1977) and pumpage has induced intrusion in Rockingham County (Atkinson and others, 1986).

New Jersey: Intrusion of sea-water is a major problem that has been monitored since 1923 (Ayers and Pustay, 1988). In 1977, the salt-water monitoring network included more than 400 wells (Schaefer, 1983). Overpumping, leaky casings, and dredging have led to intrusion in Salem, Gloucester, Cape May,

112

Middlesex, Monmouth, Ocean, and Atlantic Counties (Miller and others, 1974; Newport, 1977; Schaefer, 1983). The most significant intrusions have occurred in the areas of Sayreville, where salt-water intrusion has been observed for more than 40 years, in the Keyport-Union Beach area, where the original well field had to be abandoned in 1976, and in Cape May City (Schaefer, 1983). Even areas 20 miles or more inland, such as in Gloucester and Atlantic Counties, have experienced sea-water intrusion due to pumping and corroded well casings. At Somers Point, Atlantic County, a wedge of salt water has moved 3,000 ft inland into the Cohansey aquifer. At Artificial Island, Salem County, high pumpage at a nuclear generating plant has induced salt water (Miller and others, 1974; Newport, 1977). Since the early 1970's, well waters in the Old Bridge aquifer in the Boroughs of Keyport and Union Beach, Monmouth County, have shown high chloride concentrations, indicating sea-water intrusion. Concentrations increased from background levels of less than 5 mg/L to greater than 600 mg/L in some coastal wells (Schaefer and Walker, 1981). In the Sayreville area, the transition zone between fresh water and salt water has migrated 0.2 to 0.4 miles inland between 1977 and 1981 (Schaefer, 1983). A listing of previous salt-water studies in New Jersey was presented by Schaefer (1983).

New York: Sea-water intrusion has been documented on Long Island (Lusczynski and Swarzenski, 1966), in the Town of Southampton, the Eastport-Remsenburg-Spoenk-Westhampton area, and the North Haven-Sag Harbor area (Anderson and Berkebile, 1976). Coastal plain aquifers have been contaminated with salt water in the Port Washington area, where sea water was used in settling ponds. Long-term use of the salt water in sand and gravel pits has raised chloride levels in nearby shallow and deep wells from a normal level of less than 20 mg/L to greater than 1,000 mg/L in an area of more than two square miles (Swarzenski, 1963; Miller and others, 1974).

Heavy pumping and reduced natural recharge have caused lateral intrusion of ocean water into producing aquifers on Long Island (Newport, 1977). Sea-water intrusion in southern Nassau and southeastern Queens Counties, Long Island, occurs as one wedge of salt water in shallow glacial deposits and two more wedges in the upper and lower portions of the underlying artesian aquifer (Miller and others, 1974).

Salt-water intrusion may be occurring in two areas in the Town of Southampton, the Eastport-Remsenburg-Spoenk-Westhampton area and the North Haven-Sag Harbor area, Long Island. Salt water underlies a lense of fresh water in the area, possibly allowing salt-water intrusion both laterally and vertically. The depth to salt water in three testholes was found to be at 350, 620, and 1,060 ft below sea level, respectively. Chloride concentrations in affected wells range from 200 to 13,890 mg/L (Anderson and Berkebile, 1976).

North Carolina: Sea-water intrusion has been documented in Carteret, Pamlico, Beaufort, Hyde, Dare, and Tyrell Counties (Atkinson and others, 1986). Tidal action resulted in intrusion of saline water into ground-water sources through drainage canals on former marsh land in the Coastal Plains of eastern North Carolina. This forced abandonment of large areas of cropland in parts of Tyrell, Dare, Hyde, Beaufort,

Pamlico, and Carteret Counties (U.S. Geological Survey, 1984). Heavy pumpage could also induce recharge of salt water from streams affected by sea water resulting from tidal action and/or decreased fresh-water flow.

Oregon: Sea-water intrusion has been documented in Eugene (Lane County), North Bend (Coos County), in the Willamette Valley in the West Cascades, and in Clatsop and Tillamook Counties (Atkinson and others, 1986). Sea-water overrides fresh-water aquifers in the Horsfall Beach area as a result of winter storms (Magaritz and Luzier, 1985).

Pennsylvania: Dredging of tidal rivers and heavy pumpage from wells near tidal rivers have caused sea-water intrusion near Philadelphia in the eastern part of the state (Atkinson and others, 1986).

Rhode Island: Heavy pumpage from wells near tidal rivers has caused sea-water intrusion near Warwick in Bristol and Kent Counties and near Providence in Providence County (Atkinson and others, 1986).

South Carolina: Sea-water intrusion has occurred at several locations along the state's coast, including areas in Charleston, Beauford, and Horry Counties (Miller and others, 1977; Atkinson and others, 1986). Background concentrations of sodium and chloride in fresh ground water within the Black Creek aquifer of Horry and Georgetown Counties are less than 280 mg/L and 40 mg/L, respectively. Concentrations above that may indicate mixing of fresh water with sea water (Zack and Roberts, 1988). The most extensive encroachment at present occurs in an area that extends from the Beaufort Basin to the Cape Fear Arch (Siple, 1969). The upper zones of Eocene limestones and the sub-sea contact of Eocene and Oligocene deposits are subject to considerable salt-water encroachment, with maximum chloride concentrations of up to 8,500 ppm (Siple, 1969)

Texas: Sea-water intrusion is occurring in the Galveston, Texas City, Houston, and Beaumont-Port Arthur areas and around Corpus Christi (Atkinson and others, 1986). Along the Gulf Coast, sea water mixes with surface water in the tidal reaches of the rivers, such as in the Calcasieu River channel, Calcasieu Lake, Sabine Lake, and lower Sabine River (Krieger and others, 1957). Sea-water intrusion occurs also along the lower 36 miles of the Neches River and 3 miles of Pine Island Bayou (Port of Beaumont) during 6 months of the year. Salt-water barriers, which divert fresh water into a canal system upstream, cause this unhindered inflow of salt water; tidal action flushes the salt water back and forth below the barriers causing it to become increasingly concentrated. Deterioration of water quality is aggravated by waste disposal into the water way (Harrel, 1975). Sea-water intrusion from the Houston Ship Channel has been reported from shallow wells between Baytown and Houston. Large withdrawals of ground water (525 million gal/d), resulting in a decline of artesian pressure equivalent to a 400 ft drop of hydraulic head, has caused this intrusion over an area up to four miles wide at Baytown at a rate of several hundred feet per year (Jorgensen, 1981).

Virginia: Sea-water intrusion due to the overpumping of fresh ground water in areas of shallow saline ground water has occurred in Northhampton and Isles of Wight Counties (Atkinson and others,

114

1986) and in the areas of Newport News and Cape Charles (Newport, 1977). According to Larson (1981), historic changes in chloride content of water wells are a local rather than regional phenomena. In general, a wedge of salt water coincident with the mouth of Chesapeake Bay extends into the York-James and southern Middle Neck Peninsula, where the greatest chloride increase (175 mg/L) was measured.

Washington: Several cases of sea-water intrusion have been documented in the Puget Sound areas and along the Pacific Coast. Affected areas include Island County, the areas around Tacoma and Olympia, Kitsap County, and northern Jefferson County (Kimmel, 1963). While sea-water intrusion appears to be a more widespread problem in Island and San Juan Counties, it is of only local significance in Clallam, Jefferson, Pierce, Thurston, and Whatcom Counties (Dion and Sumioka, 1984).

3.4 Oil-Field Brine

3.4.1 Mechanism

This chapter deals with salt water that is produced with oil and gas. This water is naturally-occurring, deep-basin formation water that differs from the type discussed in chapter 3.1 more through its mechanism of mixing with fresh water than through its water chemistry. The major difference in the mixing mechanism is that formation brines unassociated with oil and gas are normally separated from fresh ground water by a transition zone of slightly to very saline water, which normally lessens the degree of natural or induced salinization. Anthropogenic contamination of fresh water by oil-or gas-field brine, in contrast, is not associated with a transition zone but instead brings concentrated brine into direct contact with fresh water. Therefore, salinization of fresh ground water by oil-and gas-field brine is often very abrupt, characterized by large increases in dissolved solids within relatively short time periods and short distances.

There are currently 25 major oil-and gas-producing states in the country (Fig. 4). In those states, more than one million holes have been drilled in search for oil and gas. According to Newport (1977), this drilling for oil and all other contamination hazards associated with the oil and gas industry are major contributors to salt-water intrusion in the inland part of the United States.

Contamination hazards associated with the oil and gas industry stem largely from the huge amount of salt water that is produced with oil and gas. In Texas, approximately 2.5 barrels of salt water were produced with every barrel of oil in 1961 (McMillion, 1965). Miller (1980) estimated that 4 barrels of salt water are produced with every barrel of oil. Others estimate ratios of salt water to brine of up to 20:1, with ratios generally increasing as the production of oil from a field decreases. In 1956, salt production at oil fields in the United States totaled 125 million tons, which is equivalent to approximately one-half of the stream discharge in the United States (Thorne and Peterson, 1967). The total amount of salt-water production increases with time, as documented by early data reported by Collins (1974) (7.7 billion barrels per year) and more recent data reported by Michie & Associates (1988) (22 billion barrels in 1986). Production data for 1963, listed state-by-state in Table 8 (Miller, 1980), identify Texas, Kansas, and Oklahoma as the three

115

Table 8. Methods of disposal of produced oil- and gas-field brine in the United States during 1963 (data from Miller, 1980).

State	Current volumes produced	Injection water flooding	Injection disposal only	Surface Disposal Lined pits	Unlined pits	Streams	Other methods	Documented brine pollution
Alabama	2,493		1,397		493	603		(B) (C)
Arizona	100				100			
Arkansas	539,132	89,082	340,734		7,444	101,871		(B) (C)
California	2,740,850	445,768	208,665	3,127	399,933	501	1,682,855	(B)
Colorado	202,194	131,500	5,000	70	65,624			(B) (C)
Florida	600		600					
Georgia (A)								
Illinois	876,712				15,132		876,712	(B) (C)
Indiana	81,797	50,960	14,724		9,600		982	(B) (C)
Kansas	5,011,400	800,000	4,200,000	1,800				(C)
Kentucky	123,287	73,973	35,616	2,740	5,480	5,480		(B) (C)
Louisiana	2,785,000	184,000	1,762,000		698,000		141,000	(B) (C)
Michigan	149,587		147,849		982		756	(C)
Mississippi	340,079	40,000	203,836	8,219	74,329		13,699	(B) (C)
Montana	50,000	10,000	31,400		8,600			(B) (C)
Nebraska	121,907	17,329	7,567		97,011			(C)
Nevada (A)								(C)
New Mexico	356,624	55,176	165,423		136,025			(B) (C)
New York (A)								(B) (C)
North Dakota	31,000	23,500			7,500			(C)
Ohio (A)								(B) (C)
Oklahoma	3,751,911	3,160,577	583,280	5,370	2,685			(B) (C)
Pennsylvania	191,780					191,780		(B) (C)
South Dakota	68				68			(C)
Texas	6,127,671	2,736,755	1,472,954		1,262,719	615,566	39,677	(B) (C)
Utah	81,634	2,981			4,862		73,790	(C)
West Virginia	115,068					115,068		(B) (C)
Wyoming (A)								(C)
Totals	23,682,022	7,821,601	9,182,173	21,326	2,796,587	1,030,869	2,829,471	

(A) Brine data not available; (B) from Miller, 1980; (C) see section 3.4.5

largest producers in the country, making up more than 50 percent of the total production. In 1986 the total volume of brine disposal at oil and gas fields in the United States amounted to approximately 60 million barrels per day. Of this, 42 million were injected into producing oil and gas formations, 17 million were injected into salt-water formations, three million were released into surface waters, and two million percolated into the ground where an underground source of drinking water is not present (primarily in the San Joaquin Basin of California) (Michie & Associates, 1988). Assuming a conservative average TDS concentration of 50,000 mg/L in those waters, the disposal equals a total salt load of 8.6 million metric tons per year deposited into surface waters and of 5.8 million tons per year percolated into the ground.

Contamination of ground water and surface water can occur where the disposal of this produced water is done in a way that allows mixing between brine and fresh water. This potential hazard is highly dependent on the disposal method. Surface disposal and brine ponding in unlined surface pits have been used widely in the past, which may have caused high potentials of fresh-water contamination in Texas, Louisiana, and California, where the largest amounts of brine were disposed of by this method (Table 8). According to Miller (1980), at least 17 oil-and gas-producing states (Table 8) have experienced water pollution from disposal of oil-field brine, with a high likelihood that contamination has taken place at one time or another in all oil-producing states.

The following sections will describe mechanisms that allow mixing of oil-and gas-field brine with fresh ground water.

3.4.1a. Surface disposal

Discharge of oil-field waters into coastal waterways, bayous, estuaries, or into inland streams, creeks, and lakes directly pollutes surface waters. Where these surface-water bodies are interconnected with ground water, ground-water pollution will also occur. Spillage from disposal and drilling pits has the potential to contaminate surface waters and leakage through the bottom of a pit has the potential of contaminating the vadose zone and ground water underlying the pit.

At earlier times with little regulation of the oil and gas industry, the often indiscriminant disposal of brines into surface waters caused severe contamination in many areas. For example, chloride content in the Green River at Mundfordville, Kentucky, increased from pre-oil-development levels of 10,600 tons in 1957 to post-oil-development levels of 305,000 tons in 1959, which is equal to a 3,000 percent increase (Krieger and Hendrickson, 1960). Indiscriminant disposal of brines onto the land surface has caused "vegetative kill" areas in many old oil and gas fields in the past, and may still be a cause of serious degradation of surface and ground water as a result of leaching. In an effort to reduce water pollution, evaporation pits were introduced as a way of brine disposal under the premise that the volume of brine could be reduced by evaporation. As a consequence, during 1961 alone, within the Ogallala outcrop of the Texas High Plains, more than 66 million barrels of brine were disposed of on the surface, primarily in unlined surface pits (Burnitt and others, 1963). This represented approximately 55 percent of that year's total brine production in the area. From a single oil field in Winkler County, Texas, more than 500,000 acre-ft of brine were disposed of between 1937 and 1957 into unlined surface pits or into a communal disposal lake (Garza and Wesselman, 1959), adding millions of tons of salt to soils and causing contamination of local water wells. In Colorado, 27 million barrels of brine were disposed of in unlined pits in 1969 (Rold, 1971). After some years of pit use, it was eventually realized that most of the disposed of brine did not evaporate but infiltrated instead into the ground through the often porous pit bottom, creating a multitude of point-contamination sources of brine. For example, in the vicinity of Cardington, Ohio, an area of 13 mi^2 was affected by brine pollution only three years after the first successful oil well had been drilled. Disposal of brine into surface pits and indiscriminant surface dumping of salt water by contract truckers was the major cause of this contamination (Lehr, 1969). As a consequence of this pollution, an entire well field had to be abandoned. Similar uncontrolled surface discharges were reported by Baker and Brendecke (1983) in Utah, where water haulers may dispose of brine into unlined trenches, surface depressions on undeveloped land, or into roadside ditches.

Release of salt from soil underlying former disposal pits can affect ground water for long times and in repeated plumes. Cyclicity of salt release from the soil is caused with every precipitation period that flushes salt from the vadose zone into the saturated zone. This causes extremely high salt concentrations under some pits even many years after their abandonment, as measured by Pettyjohn (1982) among others under a pit in Ohio (Cl of 36,000 mg/L after 8 years) and by Richter and others (1990) under a pit in Texas (Cl of 20,750 mg/L after 20 years). Even longer time periods will be needed to see salt concentrations return to background levels. Contamination of ground water caused by an unlined surface-disposal pit in southwest Arkansas covers an area of approximately one square mile and is estimated to last for another 250 years (Fryberger, 1972). The Oklahoma Water Resources Board (1975) estimated that it will take more than 100 years to flush all the salt from a 9 mi^2 area contaminated by pits in the Crescent area, Oklahoma. Disposal pits that have been used for only relatively short periods also can contaminate soil and ground water, especially if large amounts of brine have been disposed of. For example, Lehr (1969) reported contamination of ground water from two pits at Delaware, Ohio, which had been used for only 15 months, during which they had received more than 225,000 barrels of brine. As a result of this disposal, chloride concentrations in ground water increased from background levels of less than 10 mg/L to more than 35,000 mg/L. A single pit can cause ground-water deterioration on a wide scale. Rold (1971) estimated that one pit caused a 27 ppm per year salinity increase in the Severance ground-water basin of northeast Colorado. This pit had received only 200 barrels of brine per day but had been excavated into 7 ft of gravel that directly overlay the local fresh-water aquifer (Rold, 1971). The amount of fresh water lost to this kind of salinization can be substantial, such as in the case of Miller County, Arkansas, where seepage from three disposal areas contaminated approximately 60 million gallons of fresh water (Ludwig, 1972).

Often, highly saline fluids are also used in the oil and gas industry during the drilling of boreholes. These fluids are temporarily stored in pits and can cause fresh-water contamination if not disposed of properly after drilling has been completed. Murphy and Kehew (1984) estimated that approximately 360 million ft^3 of brine have been buried in shallow pits in North Dakota during pit closure, causing variable degrees of soil-water and ground-water contaminations locally.

3.4.1b. Injection wells

Injection of salt water is done either for enhanced recovery or for brine disposal. Enhanced recovery occurs in the producing formation, which typically is not a fresh-water aquifer or an aquifer containing potentially usable ground water (TDS <10,000 ppm). In some instances, especially in shallow oil fields, however, a hydraulic connection may exist between the oil pool and usable water updip. These areas (Fig. 45) are especially prone to salt-water contamination. Where this hydraulic connection does not exist, it may be created if injection pressures exceed lithostatic pressures (bottomhole pressure greater than 1 psi/ft of overburden).

Disposal of salt water may occur in any formation that is capable of accepting large amounts of fluids, with regulations concerning depths of disposal and water quality in the disposal zone varying from state to state. Where injection operations are faulty, ground-water contamination may occur where injection wells penetrate fresh-water units and salt-water units. Some of the possible mechanisms of mixing between brine and fresh water are direct injection into a fresh-water or potable-water unit, causing failure of the injection well adjacent to a potable-water zone or to a zone in hydraulic connection with potable water, intrusion of salt water into potable water as a result of increased pressure and natural leaky conditions between the units, hydraulic fracturing, or upward migration of brine along the outside of the casing (Miller, 1980). Contamination may also be associated with insufficiently plugged, abandoned boreholes that penetrate the injection zone, as discussed next.

3.4.1c. Plugged and abandoned boreholes

Exploration for oil and gas has created millions of boreholes that penetrate shallow, fresh-water aquifers and deep, saline-water aquifers. Many of these holes may have been plugged and abandoned in a condition that may allow communication between the different water types at some time (Fig. 46). Flow of brine into fresh water along abandoned boreholes can occur where the brine unit or the fresh-water unit are not sealed and where the hydraulic head of the brine unit is higher than the hydraulic head of the fresh-water unit. Higher hydraulic heads in brine units may be natural or man-induced, the latter either by injection into the brine unit or by the lowering of heads in the fresh-water unit. Most critical are those wells that were abandoned without sufficient plugs and casing in the early stages of oil and gas exploration and many relatively shallow (up to several hundred feet deep) seismic shotholes. But even boreholes that

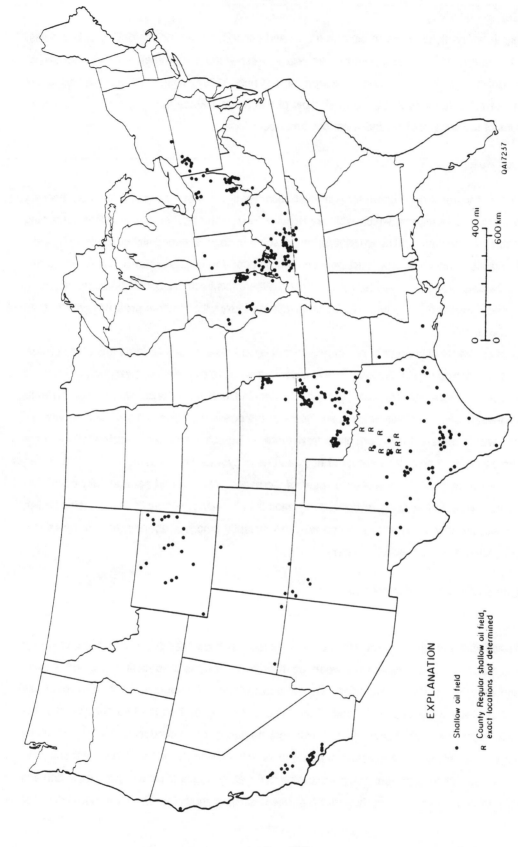

EXPLANATION

• Shallow oil field

R County Regular shallow oil field,
   exact locations not determined

Figure 45. Map of shallow oil fields in the United States (from Ball Associates, Ltd., 1965).

QAI7257

400 mi

600 km

120

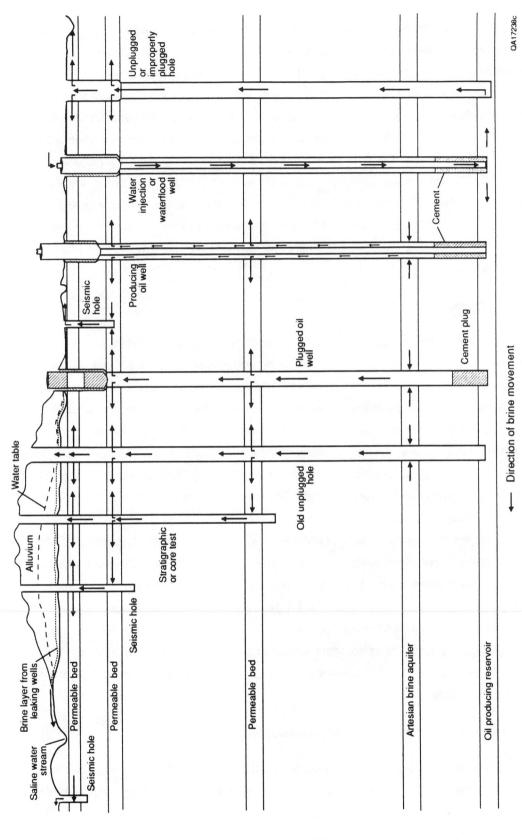

Figure 46. Schematic diagram illustrating possible communication scenarios between deep saline aquifers and shallow fresh-water units through boreholes (from Burnitt, 1963).

121

have been plugged and abandoned in a condition sufficient to protect potable water at the time of abandonment may present potential contamination hazards in the future, as the life of steel casings, depending on the quality of the cement job, may only last for 5, 10, or 20 years (Rold, 1971).

Inadequately plugged and abandoned boreholes may provide pathways for upward migration of brines where natural potentials of upward flow exist. As such, they act similar to faults and fractures that allow vertical migration of saline waters. In the absence of a flow barrier, hydraulic heads in deep brine formations higher than the water table or potentiometric surface of an overlying fresh-water aquifer will cause mixing of brine with fresh ground water. This mixing is chemically similar to the mixing of oil-field brine with fresh ground water. The history of salinization, geologic conditions, and the local or regional character of the occurrence of saline ground water are often important parameters to distinguish between this natural mixing of saline formation water with fresh ground water and the mixing of disposed of oil- or gas-field brine with fresh water.

3.4.2. Oil-Field Brine Chemistry

Overall salinity, concentration ranges of individual constituents, and the types of chemical constituents vary much more in oil and gas brines than in previously discussed brines, that is, halite-solution brine (chapter 3.2) and sea-water intrusion (chapter 3.3). As seen in most natural brines, a strong correlation exists between sodium and chloride concentrations (Fig. 47). The Na/Cl weight ratio is typically less than approximately 0.60, often reflecting high concentrations of other cations, especially calcium (Fig. 47). The relatively large scatter on plots of major cations and anions versus chloride (Fig. 47), especially of SO_4 versus Cl, suggests significant differences between oil-field brines from different areas or reservoirs. These differences can possibly be used to identify the source of contamination in areas where production is from more than one reservoir.

The most common constituents dissolved in oil and gas brines are the major cations sodium, calcium, magnesium, and potassium, and the major anions chloride, sulfate, and bicarbonate. Other constituents fall into the general ranges of approximately 100 ppm for strontium, 1 to 100 ppm for Al, B, Ba, Fe, and Li, a few parts per billion in most waters for Cr, Cu, Mn, Ni, Sn, Ti, and Zr, and a few parts per billion in some waters for Be, Co, Pb, V, W, and Zn (Rittenhouse and others, 1969). With respect to sea water, Cr, Li, Mn, Si, and Sr are commonly more than twice as abundant, whereas Cu, K, Ni, and Sn are commonly less than half as abundant. In some brines, certain elements may be at concentrations high enough to make extraction economically feasible.

3.4.3. Examples of Geochemical Studies of Oil- and Gas-Field Brine Pollution

As is the case with most salinization sources, identification of oil- and gas-field brine contamination is easy as long as it is the only suspected or possible source of salt water. Background chloride

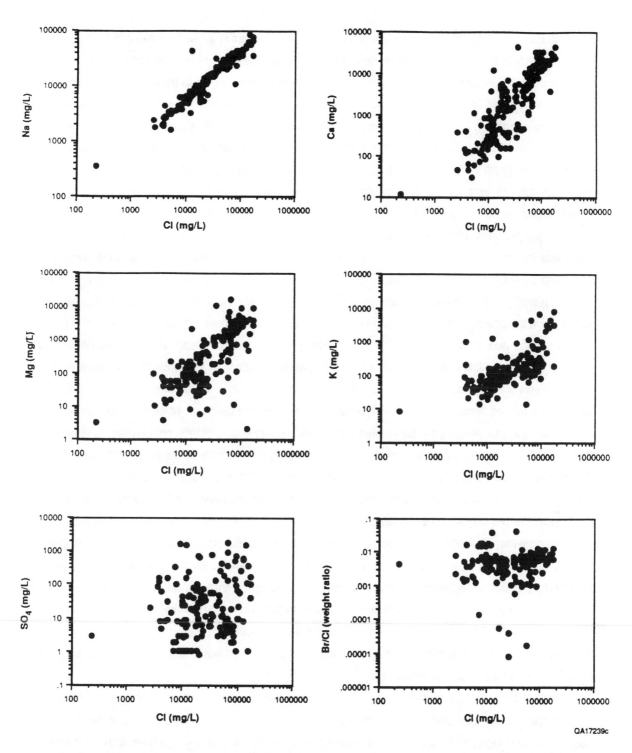

Figure 47. Bivariate plots of major ions and Br/Cl ratios versus chloride for oil-field brines from California (data from Gullikson and others, 1961), Texas (data from Kreitler and others, 1988), West Virginia (data from Hoskins, 1947), and Kentucky (data from McGrain and Thomas, 1951). Large scatter indicates a high variability in chemical composition.

concentrations in most ground waters are in the milligrams per liter to a few hundred milligrams per liter range, which is in big contrast to chloride concentrations of several tens of thousands of milligrams per liter found in most produced oil and gas waters. For example, a chloride increase from less than 100 mg/L to contamination levels of greater than 1,000 mg/L indicated contamination caused by an unlined blow-down pit used during drilling of a gas well in northeast Ohio (Knuth and others, 1990). Kalkhoff (1986) used a cut-off value of 50 mg/L in a study of ground-water contamination in parts of Mississippi, above which brine contamination by oil and gas activities was indicated. Contamination by inadequately designed disposal wells has caused increases of chloride concentrations in ground water and springs in the Upper Big Pitman Basin of Kentucky from baseline values of 4 to 43 mg/L to contamination levels greater than 50,000 mg/L (Hopkins, 1963). In the Upper Green River Basin, also in Kentucky, contamination by oil-field brine was indicated by an increase from background levels of less than 10 mg/L to levels exceeding 1,000 mg/L (Krieger and Hendrickson, 1960). Monitoring of chloride levels in observation wells at oil fields can easily detect sudden contamination caused by salt-water spills and water quality improvement subsequent to spills or disposal-pit closures (Fig. 48). Richter and others (1990) determined soil-chloride concentrations and dissolved chloride in water underlying former brine-disposal pits for identification of contamination potentials associated with abandoned pits. Comparison of soil-chloride concentrations from areas outside the pit (background levels) with those under the pit, and knowledge of the amount and salinity of disposed brine, allowed determination of the salt percentage that had been flushed out and the salt percentage that remained to be flushed out. In one example, soil-chloride concentrations were up to three orders of magnitude higher under the pit (5.8 mg/cm^3) than outside of the pit (0.007mg/cm^3) even though the pit had been abandoned more than 20 years ago (Fig. 49). Considering the amount of brine disposed of into the pit system (100,000 barrels), Richter and others (1990) estimated that approximately four percent of the original brine remained in the soil. Although this percentage appears to be relatively low, ground-water chlorinity was still very high under the pit (20,750 mg/L). Ground-water contamination spread from this pit at least 0.5 mile downgradient, as evidenced by a chloride concentration of 12,190 mg/L in a shallow testhole sample. Besides chloride concentrations, other constituents are often used to identify oil- and gas-field brine contamination. In the previous example, bivariate plots of major cations versus chloride identified disposal pits and eliminated local, saline ground water as the source of salinization (Fig. 50), which is indicated by a mixing trend between pit water and testhole waters that does not coincide with the local ground-water trend.

Krieger and Hendrickson (1960) used Piper plots to demonstrate mixing between oil-field brine and fresh water, which was suggested from high chloride concentrations. On the Piper diagram, the contamination is indicated by a straight-line mixing trend from a Ca-Mg-HCO$_3$ type fresh water to the Na-Cl type oil-field brine. Williams and Bayne (1946) used bar graphs to display differences in mixtures of fresh water and saline formation water from mixtures of fresh water with oil-field brine in Kansas. Mixtures between oil-field brines and fresh water were reflected in relatively low percentages of magnesium and

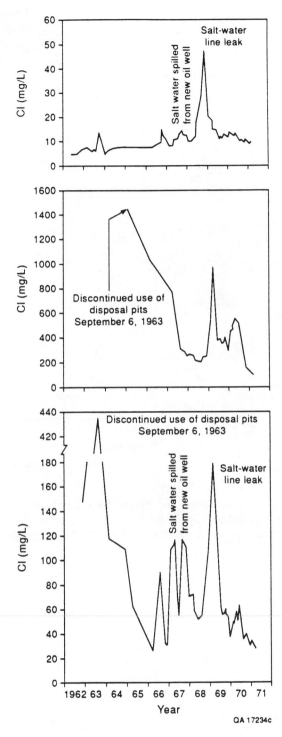

Figure 48. Monitoring of chloride concentrations in ground water as an aid in identifying salt-water contamination (from Miller and others, 1977).

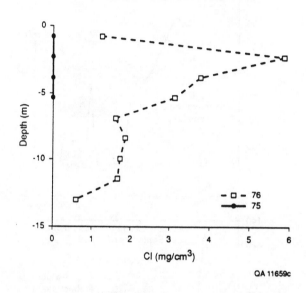

QA 11659c

Figure 49. Relationship between soil chloride and depth inside (testhole No. 76) and outside (testhole No. 75) an abandoned brine-disposal pit. High chloride concentrations in soil underneath the pit indicate that residual salt had not been flushed out since abandonment (from Richter and others, 1990).

126

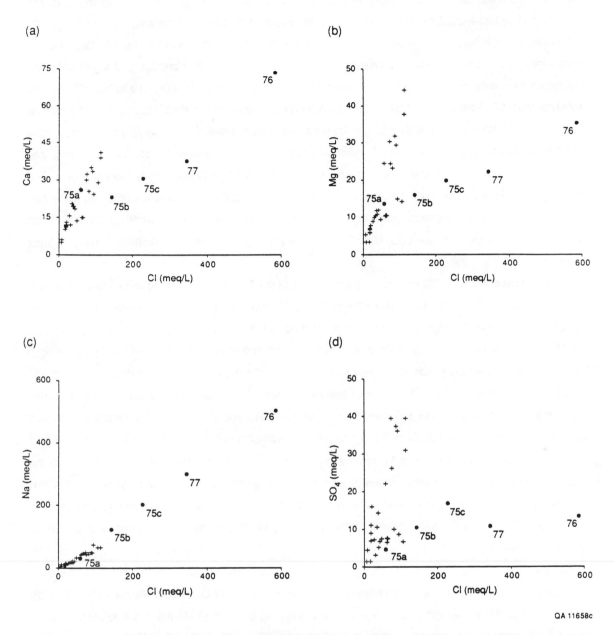

QA 11658c

Figure 50. Bivariate plots of major ions versus chloride from water-supply wells (crosses) and test holes (dots) in the vicinity of an abandoned brine-disposal pit (testhole No. 76). Test hole samples plot intermediate between the pit sample and local ground water, indicating that the salinity is derived from the pit and not from local saline ground water that affects water-supply wells (from Richter and others, 1990).

sulfate, whereas mixtures of fresh water with saline formation water contained higher magnesium and sulfate percentages. Burnitt and others (1963) used a combination of bar graphs, data posting, and Stiff diagrams to graphically depict changes in ground water quality in Ogallala outcrop areas of Texas (Fig. 51). The changes, resulting from oil-field contamination, were (1) an increase in Na, Cl, Ca, and Mg concentrations, (2) base exchange between Na in the contaminating brine and Ca and Mg in soils and caliche, causing a decrease in the Na/(Ca+Mg) ratio, and (3) low ratios of HCO_3/Cl and SO_4/Cl. Salinity diagrams (Fig. 17) were used by Nativ (1988) to differentiate between this salinization caused by oil-field brine from salinization caused by evaporation from a shallow water table. The former is typically associated with Na-Cl facies, whereas the latter is typically associated with Na-SO_4 facies. Stiff diagrams and geographic mapping of major chemical parameters were also used by Levings (1984) in a study of oil-field pollution in the East Poplar field, Montana. Elevated TDS, Na, and Cl concentrations reflected the extent of ground-water movement away from the pollution site on isocontour maps, whereas Stiff diagrams illustrated the change from low-TDS, Na-HCO_3 waters (background levels) to high-TDS, Na-Cl waters (produced oil-field waters).

Leonard and Ward (1962) were among the first to use the Na/Cl ratio to distinguish oil-field brine from halite-solution brine in Oklahoma. One type of brine, derived from salt springs in western Oklahoma, typically shows a Na/Cl weight ratio in the range of 0.63 to 0.65, which suggests that nearly pure halite (Na/Cl weight ratio of 0.648) is the source of sodium and chloride in those brines. Another type of brine in the same area consistently has Na/Cl weight ratios less than 0.60, and the ratio decreases with the increase in chlorinity (Fig. 25). This type of brine was derived from oil-field operations. Based on this difference in ratios between the two potential sources, Leonard and Ward (1962) determined that halite solution accounted for the salinity in the Cimarron River samples (#5 and #6, Fig. 25), whereas oil-field brines accounted for all or most of the salinity in samples from the Little River (#1 through #4) and the Arkansas River (#7 and #8). Surface-water degradation of streams in Kansas was studied by Gogel (1981) using the same technique. Samples from the Ninnescah River alluvium ranged from 0.65 to 0.67 in Na/Cl weight ratios, indicating that halite solution in the underlying Wellington aquifer is the source of salinity. In contrast, very low Na/Cl ratios, ranging from 0.54 to 0.28, along Salt Creek, suggest oil-field contamination. The Na/Cl ratio was also used by the Oklahoma Water Resources Board (1975) to distinguish halite-solution brine from oil-field contamination in parts of Oklahoma. Ratios of Na/Cl (0.38) identified oil-field brine as the source and chloride mapping (isochlors) identified the location of point sources as well as the extent of contaminant plumes (Fig. 52). Kalkhoff (1986) used the minor constituents bromide and strontium and the Na/Cl ratio to support his conclusions of oil-field pollution in the example noted above. In this case, a Na/Cl ratio less than 0.6 was considered indicative of oil-field pollution.

Whittemore (1984) pointed out that oil-field brines in Kansas usually have higher Ca/Cl and Mg/Cl ratios and lower SO_4/Cl ratios than halite-solution brines. The same relationship was found for halite-

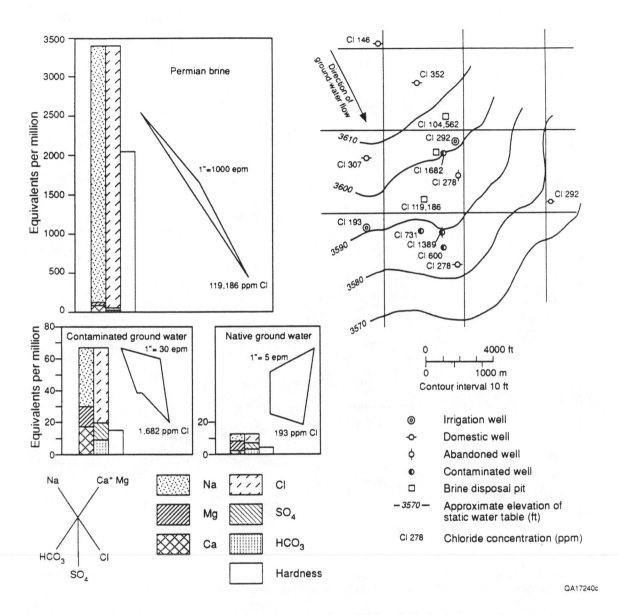

Figure 51. Identification of salinization sources using bar graphs, Stiff diagrams, and contouring of constituent concentrations onto maps (from Burnitt and others, 1963).

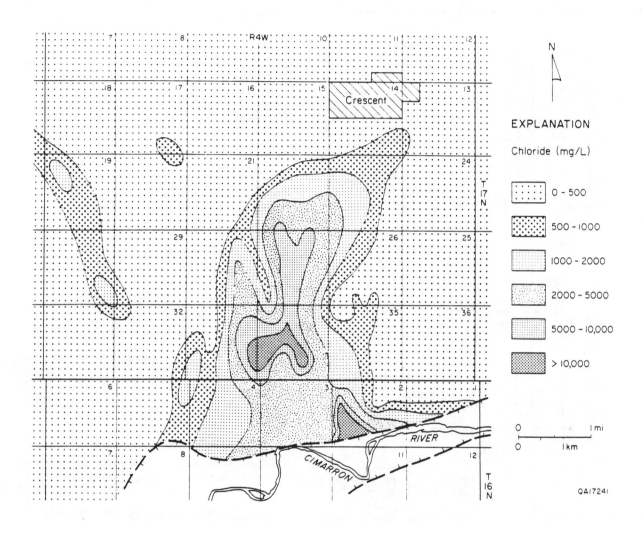

Figure 52. Mapping of point source of salinity and the resulting salt-water plume through contouring of chloride concentrations from water wells (Oklahoma Water Resources Board, 1975).

dissolution brines and deep-basin brines in Texas by Richter and Kreitler (1986a,b). However, these ratios appear to work as tracers best when little chemical reactions occur after mixing of the respective salt-water source and fresh water. To avoid this change of chemical constituent ratios by mechanisms other than mixing or dilution, Whittemore and Pollock (1979) suggested the use of minor chemical constituents, such as bromide, iodide, and lithium, that are relatively conservative in solution. Of those, bromide concentrations and Br/Cl weight ratios are the most widely used tracers because bromide is similarly conservative as is chloride, and because a significant difference in Br/Cl ratios between most oil-field waters and halite-solution brines could be established (for example, Whittemore and Pollock, 1979; Whittemore, 1984; Richter and Kreitler, 1986a,b). In Kansas, oil-field brines typically have Br/Cl ratios greater than 10×10^{-4}, whereas halite-solution brines typically have Br/Cl ratios less than 10×10^{-4} (Whittemore, 1984). This difference can be used to calculate mixing curves between fresh water composition and the range of values for either endmember (Fig. 26). By superimposing values from testholes in the Smoky Hill River area, Whittemore (1984) was able to show that halite solution is the dominant mechanism of salinization in that area (Fig. 26a). In contrast, in the Blood Orchard area south of Wichita all of the observation-well samples indicate mixing of fresh water with oil-field brines (Fig. 26b). Note in figure 26b that the sample obtained from the Arkansas River suggests halite solution; this chemical signature was derived from areas upstream from the Blood Orchard area. Also, some testhole samples plot between the two mixing fields, suggesting the mixing of fresh water with both halite-solution brine and oil-field brine. The Br/Cl ratio can also be used to identify mixtures derived from different oil-field brines as long as ratio ranges do not overlap in respective brine endmembers.

In addition to the ratios discussed above, Whittemore and Pollock (1979) pointed out the usefulness of I/Cl ratios in differentiating between oil-field brines and halite-solution brines. Halite-solution brines typically have I/Cl weight ratios less than 1×10^{-5}, whereas oil-field brines in Kansas have ratios greater than 2×10^{-5} (Fig. 31). A similar relationship in I/Cl ratios between halite-solution brine and deep-basin brine was found by Richter and Kreitler (1986a,b) for samples from salt springs and shallow test holes in north-central Texas (Fig. 28). In the Anadarko Basin, brines in Pennsylvanian and Mississippian strata are characterized by some of the highest iodide concentrations ever recorded in natural brines, with iodide concentrations of up to 1,400 mg/L (Collins, 1969). These high contents of iodide may be due to concentration by shallow-water fauna and flora and additional diagenetic concentration. In brines being more or less equal in iodide concentration, boron appears to be a good tracer to distinguish Pennsylvanian brines with low boron contents from Mississippian brines with high boron contents (Collins, 1969).

Novak and Eckstein (1988) used discriminant analysis and modified graphical techniques in combination with selected ionic ratios in a study of ground-water quality deterioration in northeastern Ohio to determine the relative importance of oil and gas brines versus road salt. By applying the ratios of Ca/Mg, Na/Ca, Na/Mg, Na/Cl, K/Cl, K/Na, Mg/K, Ca/K, Cl/Mg, Cl/Ca, and (Ca+Mg)/(Na+K), not only was it possible to identify brine contamination but it was also possible to identify the stratigraphic origin of the brine source. Other chemical and isotopic tracers of brine sources, such as bromide, strontium, oxygen-18, deuterium, carbon-13, and sulfur-34, were not considered by Novak and Eckstein (1988) because emphasis was put on easy and inexpensive analytical techniques. Depending on the number of chemical analyses available, Novak and Eckstein (1988) suggested two methods of investigation. When only a few samples were available, graphic representation of ratios on modified Piper, Schoeller, or Stiff diagrams appeared appropriate, whereas discriminant analysis proved valuable when a large amount of analyses had to be handled.

Morton (1986) identified brine contamination of fresh ground water by oil and gas operations in east-central Oklahoma using the following major criteria: (1) Cl greater or equal to 400 mg/L, (2) Br greater or equal to 2 mg/L, and (3) (Na+Cl)/TDS greater or equal to 0.64. Supporting evidence for brine contamination was provided by (4) Li/Br less or equal to 0.01 for Cl greater or equal 400 mg/L, (5) Na/Cl approximately 0.46, (6) Na/Br approximately 92, and (7) Br/Cl approximately 4.8×10^{-3}. These criteria were established by Morton (1986) through a series of constituent and constituent-ratio plots. On plots of Na/Br versus Br (Fig. 53), of Na/Cl versus Br, and of Br/Cl versus Cl, a bromide concentration of greater than 2 mg/L appears to indicate less scatter and ion ratios that are similar to or are typical for brines. A bromide value of 2 mg/L was measured in samples with chloride ranging from 250 to 500 mg/L. A value of 400 mg/L chloride was selected by Morton (1986) as the brine index because the mean Br/Cl ratio of 4.8×10^{-3} intersects the bromide line of 2 mg/L at that concentration. This value of 400 mg/L chloride also appears to define the cutoff between large and small scatters on a Na/Cl versus Cl plot (Fig. 53). Accepting the brine index of 400 mg/L chloride, a (Na+Cl)/TDS index of ≥ 0.64 is suggested. The (Na+Cl)/TDS index and the 400-mg/L-chloride index can be used to evaluate available data that don't include dissolved bromide as an analyzed parameter, as is most often the case.

Mast (1982) suggested the use of mixing diagrams (Fig. 32) for differentiating between mixing of fresh water with naturally saline ground water and mixing of fresh water with oil-field brine in Kansas. This technique is discussed in chapter 3.2 on halite solution.

3.4.4. Significant Parameters

Oil-field brines have some of the highest Br/Cl ratios found in natural salt waters. Ratios typically are greater than approximately 10×10^{-4} in oil-field brines and less than 10×10^{-4} in halite-solution brines (Whittemore and Pollock, 1979; Whittemore, 1984; Richter and Kreitler, 1986a,b). Ratio differences between these two potential endmembers of mixing with fresh water are generally big enough to allow differentiation of the respective source in brackish water down to chloride concentrations of a few hundreds of milligrams per liter, although identification is best at high concentrations.

132

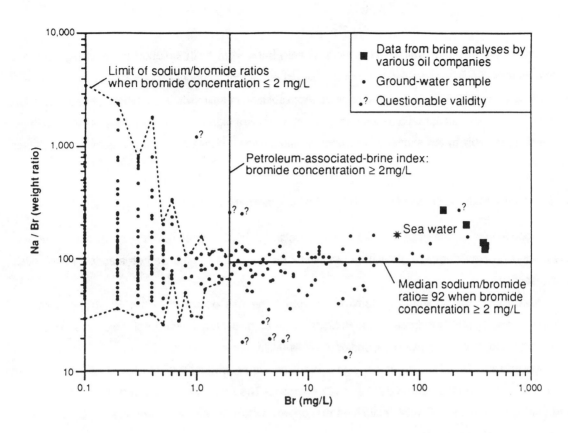

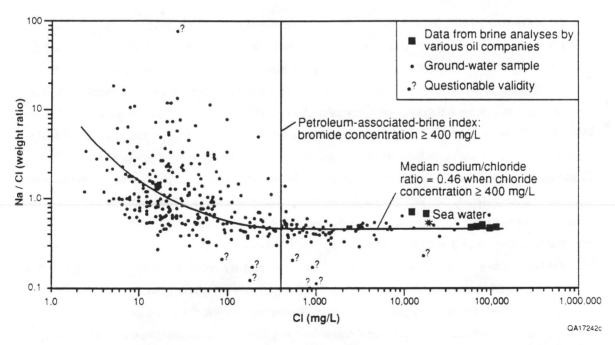

Figure 53. Determination of geochemical constituent criteria that indicate brine contamination of fresh ground water in parts of Oklahoma. Through a series of constituent plots, background values and mixing values can be established, based on theoretical mixing relationships between fresh water and brines (from Morton, 1986).

133

The ratio of Na/Cl works well to distinguish oil-field brine from halite-solution brine at high chloride concentrations. Sodium and chloride occur in halite at equal molar concentrations (Na/Cl molar ratio = 1; Na/Cl weight ratio = 0.648). Brines that originate from solution of halite within a shallow ground-water flow system will exhibit a similar ratio as long as concentrations are high enough so that the Na/Cl ratio is not affected appreciably by ion exchange reactions. In most oil-field brines molar Na/Cl ratios are much less than one.

3.4.5. State-by-State Summary of Oil-and Gas-Field Contamination

Production of oil and gas once was or is now occurring in parts of 25 states (Fig. 4). Associated with this production are different degrees of ground-water problems, as summarized on a state-by-state basis in this section.

Alabama: Ground-water contamination from brine has occurred in the Pollard, Gilbertown, Citronelle, and South Carlton oil fields. Principal sources of contamination in oil fields are (a) unlined disposal pits, (b) leaks from pipelines, and/or (c) spills (Miller and others, 1977).

Arkansas: The disposal of brine and brackish water produced with oil and gas has caused deterioration of several ground-water basins in Arkansas. The Arkansas Department of Pollution Control and Ecology (1984; Morris, 1988) estimated that approximately 20,000 acres are affected by petroleum-industry waste and salt water.

Ground-water salinization has been reported in Independence, White, Woodruff, Chicot, Ashley, and Union Counties (Atkinson and others, 1986). In Miller County, brine-polluted ground water (Cl>40,000 mg/L) can be found in river alluvium along the Red River west of Garland City. A faulty disposal well and disposal of brine into a surface pit accounted for a brine plume covering one square mile. It is estimated that this plume will extend within the next 250 years to 4.5 mi^2 before all salt is flushed into the river. Two other polluted areas and four possibly polluted areas have been found in this county (Fryberger, 1972) where more than 60 million gallons of fresh water have been contaminated by brine disposed into surface pits (Ludwig, 1972). Disposal of oil-field brines also contaminated many streams and associated ground waters. Streams that have been most affected are in the southern part of the state and include Bodcaw, Cornie, Smackover, and Lapile Creek, Bayous Dorcheat, and the southern reaches of the Ouachita River (Scalf and others, 1973).

Oil-field brines in southern Arkansas are characterized by relatively high calcium (30,000–50,000 mg/L) and bromine (1,500 to greater than 3,000 mg/L) concentrations (Collins, 1974).

Colorado: Disposal of brine and inadequate abandonment of exploration and production wells has led to numerous occurrences of ground-and surface-water pollution. Uncased exploratory oil wells in the White River area caused an increase in the river's salt load from 9.9 million lb/year to as high as 52.9 million lb/year (Feast, 1984). For example, a 1,837 ft deep oil test drilled in 1915 was allowed to discharge approximately 52,000 metric tons of TDS per year (TDS of 19,200 mg/L at a discharge rate of 1,350 gpm) over a 53-year period in unrestricted flow to the White River in western Colorado. Subsequent plugging of this well caused other nonflowing wells in the area to flow and created saline seeps in the vicinity of these wells (van der Leeden and others, 1975). During 1969, approximately 27,000,000 billion barrels of salt water were disposed of in surface pits, affecting entire ground-water basins. For example, the South Platte River Basin exhibits annual increases in TDS, with total concentrations currently ranging from 500 to 4,000 ppm. One single disposal pit in the Severence Basin at Weld County contributes approximately 27 ppm TDS to the entire ground-water basin (Rold, 1971).

Georgia: An abandoned oil-test well in Glynn County allowed upward flow of saline water from saline zones below 2,000 ft into several fresh-water zones between 610 and 920 ft. Chloride concentrations vary from 40,000 ppm in sands at 4,150 ft depth to as little as 18 ppm at 1,500 ft, resulting in a discharge-water concentration of 7,780 ppm at land surface (Wait and McCollum, 1963).

Indiana: An abandoned, unplugged exploration borehole at Terre Haute polluted three high-capacity water wells 2,000 ft away with saline water containing 8,600 ppm of chloride. Static water level in the deep hole was 28 ft above the static water level of the shallow aquifer (Gass and others, 1977).

Kansas: Many areas in the state have been affected by oil and gas activities. Past disposal of oil-field brine still yields salt water to surface water in some oil-field areas (Krieger and others, 1957).

Oil-field brine is a major source of saline water in the Walnut River Basin, contributing perhaps 40 percent of the TDS load downstream from the confluence of the Walnut and Whitewater Rivers (Leonard, 1964). Shallow ground water in part of the Little Arkansas River Basin in south-central Kansas has been contaminated by oil-field brines and municipal wastes. Oil-field brines enter shallow ground water through corroded or improperly cased wells, by upwelling of injected brine, and by leakage from disposal ponds (Leonard and Kleinschmidt, 1976). Oil-field brine also mixes with fresh ground water and with saline formation water along the eastern side of the Elm Creek Valley in Barber County (Williams and Bayne, 1946) and around Salina (Atkinson and others, 1986).

Kentucky: Salt-water problems in Kentucky are generally associated with the oil and gas industry and not with overpumping and inflow of natural saline ground water, although most of the state is underlain by shallow saline ground water (Atkinson and others, 1986).

Oil-field brines have contaminated shallow ground water in the Upper Big Pitman Creek Basin. Contamination started with the practice of brine disposal into ponds, sinks, and local drainage ways and continued with brine injection into the shallow subsurface (175 ft below land surface) due to the presence of inadequately abandoned test wells. Chloride concentration increased from 60 ppm prior to oil production to as high as 51,000 ppm after oil production. Secondary-recovery operations may increase the extent of contamination if abandoned wells are not plugged (Hopkins, 1963).

Brine disposal at oil fields have contaminated water wells and springs in the upper Green River Basin, especially in Green and Taylor Counties. Before oil production (August 1958), river water was of the

135

Ca-Mg-HCO$_3$ type with an average chloride concentration of less than 10 ppm. With the development of the oil field, the water changed to a Na-Cl type with chloride concentrations frequently exceeding 1,000 ppm. Chloride discharge of the Green River at Munfordville increased from 10,600 tons in 1957 to 305,000 tons in 1959. An increased chloride load is reported as far as 100 miles downstream from the areas of heaviest brine production (Krieger and Hendrickson, 1960).

Upward flow of oil-field brine from disposal zones in the Louisville Limestone through abandoned oil and gas wells has contaminated shallow ground water in the state; chloride concentration increased locally from 60 mg/L prior to oil production to 51,000 mg/L after oil production (Gass and others, 1977).

Louisiana: Oil and gas activities in salt-dome areas in the southern and northern parts of the state are potential sources of salinization. (Atkinson and others, 1986). Oil-and gas-field operations contribute large amounts of brine to surface-water bodies along the Louisiana coast, as state regulations allow disposal of brine into naturally saline or otherwise unusable water. Since 1938, millions of barrels of brine have been disposed of into lakes, canals, bayous, and marshes in Lafourche Parish, contaminating sediments and surface waters and destroying vegetation (Haque, 1989). In 1986, approximately 2.6 million barrels of oil-field brine, which is equivalent to 70 percent of the state's total brine production, were being discharged into surface water at 698 stations along the coast (Van Sickle and Groat, 1990). Lately this practice has raised new concern because of occasionally high concentrations of naturally-occurring radioactive materials in oil-field brines.

Disposal of drilling and production wastes has contaminated surface and shallow ground water in Vermilion Parish, where these wastes had been disposed of by injection, unlined surface impoundments, land application, landfill, burial, and marsh reclamation (Subra, 1990).

Michigan: Hundreds of millions of gallons of highly mineralized water have been leaking through abandoned boreholes during the past 80 to 100 years, creating widespread salinization throughout the state (Gass and others, 1977). The Michigan Department of Natural Resources (1982) reported contamination cases caused by oil-and gas-field activities in 19 counties during 1979.

Mississippi: The first successful oil well was drilled in Yazoo County in 1939. Since then, several thousand wells have been drilled throughout the state, making brine disposal one of the major sources of salt-water contamination in the state. According to an inventory by Miller and others (1977), in 13 of 25 counties in which brine-disposal wells are located, contamination of ground water due to oil-field brines has been reported. Saline springs and seeps resulting from the disposal of brine into unlined surface pits have been encountered in Wayne, Wilkinson, and Yazoo Counties. Contamination of surface and shallow ground water by oil-field brine has also been reported in the Brookhaven, Baxterville, Pistol Ridge, Little Creek, and Tinsley oil fields as well as at numerous areas in Adams County (Kalkhoff, 1986), where chloride background levels in uncontaminated fresh water are less than 20 mg/L. Affected surface waters include Tallahala Creek, Leaf River, Chickasawhay River, Eucutta Creek, Yellow Creek, and Hortons Mill Creek (Shows and others, 1966; Baughman and McCarty, 1974).

Montana: Locally, deterioration of ground-water quality has occurred from leaky wells and brine disposal. Major areas of oil production are located in the Powder River Basin, the Williston Basin, and the Sweetgrass Arch. During the period of 1950–1970, ten seismic companies drilled more than 300,000 seismic holes in Montana. There is concern that holes drilled during oil and gas activities may allow crossformational flow between multiple aquifers as a result of head differentials in inadequately plugged wells and test holes (van der Leeden and others, 1975).

Disposal of oil-field brine in surface pits and wells or leaky wells and pipelines have caused contamination of surface water and shallow ground water in the East Poplar oil field, Roosevelt County. Chloride concentrations as high as 45,000 mg/L were measured in testholes 45 ft deep. The Poplar River exhibits a chloride increase from 20 mg/L to 880 mg/L in the area, associated with a change in water facies from a Na-HCO$_3$ type to a Na-Cl type (Levings, 1984).

Nevada: The only oil-producing area is Eagle Springs field in Railroad Valley of east-central Nevada (Van Denburgh and Rush, 1974). Disposal of brine in that field occurred through ponds, from which the water infiltrated the valley-fill alluvium. Since oil production began in 1954, about 500 acre-ft (through 1971) of brine containing 25,000 to 30,000 mg/L TDS were disposed of by this method. This amount is equal to 17,000 to 20,000 tons of salt. There are indications that shallow ground water is affected by the percolating brine, as suggested by an increase in chloride concentrations from background levels smaller than 30 mg/L to contamination levels of 66 mg/L in a 79-ft domestic well (Van Denburgh and Rush, 1974).

New Mexico: Oil- and gas-field operations have caused salinity increases in the San Juan River valley-fill aquifer and in the Pecos River valley-fill aquifer (Ong, 1988). Instances of ground water salinization in the southern part of the state are suspected to have been caused by disposal of oil-field brines (U.S. Geological Survey, 1984), such as in Lea County, where a leaky brine pit contaminated fresh ground water (McMillion, 1970).

New York: Oil and gas seeps have been known to occur in western New York since historic time, as documented by accounts about Indians collecting oil from the Seneca Oil spring near Cuba. Other sites of oil springs are at Freedom, Allegany County, and around Canandaigua Lake. This discharge of gas or oil at land surface is probably too small to have contaminated water supplies; an exception to this might be Oil Creek near Cuba. Oil production is concentrated in Allegany and Cattaraugus Counties and to some extend in Steuben and Chautauqua Counties. These counties have the greatest potential of oil-well pollution. Among the areas of oil-field pollution are the valleys of Chipmunk and Knight Creeks and Genesee River, where surface-water chlorinities of up to 4,000 mg/L have been recorded (Crain, 1969). In 1973, there were approximately 5,400 operating oil and gas wells in the state (Miller and others, 1974). Some domestic wells in Chemung County appear to have been contaminated by mineralized water that entered shallow aquifer units through old gas wells, as indicated by an increase in chloride concentrations from background values of less than 10 mg/L to several hundreds of mg/L (Miller and others, 1974). Randall (1972; Miller and others, 1974) suggested similar conditions for the Susquehanna River Basin. In

137

the Jamestown area, abandoned oil and gas wells, which had been drilled through beds of halite, are suspected to allow vertical migration of salt water into shallow fresh-water aquifers (Crain, 1966; in Miller and others, 1974).

North Dakota: Oil- and gas-field activities, which started in the early 1950's, contribute to the salinization problem in the state. Between 1951 and 1984, an estimated 9,000 wells were drilled in the state. Associated with drilling is the use of highly saline drilling fluid, which is commonly buried in pits and trenches after drilling has ceased. Murphy and Kehew (1984) estimated that approximately 360 million cubic feet of drilling have been buried that way in the state.

Ohio: In mid-1979, roughly 35,000 active oil and gas wells produced approximately 40,000 barrels of brine per day (Templeton and Associates, 1980). These brines were produced nearly exclusively from the Clinton, Trempealean, and Berea Formations. Disposal of these brines has caused major problems in the state (Atkinson and others, 1986).

Water pollution associated with oil-field brines has been documented in Morrow, Delaware, and Medina townships (Pettyjohn, 1971). Approximately 13 mi^2 of land have been affected by oil-field brine in Morrow County after only 3 years of oil production (Lehr, 1969). Contaminated ground water underlying the affected areas appears to move like a slug and does not mix extensively with fresh water. Major causes of contamination are disposal of brine into surface pits and the indiscriminate dumping of salt water by contract truckers. Two disposal pits at Delaware, Delaware County, have contaminated ground water, affecting approximately 20 acres of land along the Olentangy River (Lehr, 1969).

Ground water has been and still is being contaminated by abandoned brine-disposal pits along the Olentangy River in central Ohio. Leaching of salt from soil underneath the pits occurs with each precipitation event, creating repeated plumes of salt water in the area. Chloride concentrations in the plume often exceed the concentration of the original brine disposed of in the pits (Cl=35,000 mg/L) due to brine evaporation prior to infiltration. The process of leaching out of soil salts appears to continue for many years after ponds have been closed and reclaimed (Pettyjohn, 1982).

Leaky brine pits, insufficient surface casing, and poor cementing have caused fresh-water contamination in Perry Township, Lake County (Novak and Eckstein, 1988).

Oil brine is spread onto roads as an anti-dust agent in 77 townships of Ohio, resulting in elevated concentrations of heavy metals in soils (Kalka and others, 1989). Breen and others (1985) estimated that of the 38,000 barrels of brine produced every day in Ohio, approximately 75 percent are disposed of by road spreading, evaporation ponding, and illicit dumping. Contamination of surface and ground water from brine spreading deviates from contamination caused by road salting (rock salt or salt-solution brine), although most of the oil-field brine constituents appear to be absorbed in the ground within a few tens of ft from the roads (Bair and Digel, 1990). Periodic application during winter and summer months can lead to high chloride concentrations in ground water adjacent to highways.

Oklahoma: Oil-field brines constitute problems in the Cimarron Terrace from Woods County southeast to Logan County (Atkinson and others, 1986).

Many streams have been contaminated by salt water from the disposal of oil-field brines, leading to pollution of shallow ground water recharged by these streams. The Arkansas, Canadian, Cimarron, and Red Rivers have been affected by salt-water. Oil-field brine pollution has been reported from Garvin, Pontotoc, Seminole, Oklahoma, Pottawatomie, Lincoln, Okfuskie, and Creek Counties. In 1970, approximately 15,000 salt-water injection wells existed in Oklahoma (Scalf and others, 1973).

Salt springs and seeps issuing from underlying salt beds increase the salinity of the Cimarron River near Mocane. The salinity of the river water is increased further in the lower reaches by oil-field brines. Oil-field brine also causes salinization of the North Canadian River downstream from Oklahoma City, amounting to TDS concentrations at times exceeding 15,000 ppm (Krieger and others, 1957).

Salt-spring brines in western Oklahoma can be distinguished from oil-field brines using the ratio of sodium over chloride. Spring brines typically exhibit Na/Cl ratios greater than 0.60 (weight ratios), suggesting halite dissolution to be the source of the salt water. Oil-field brines, in contrast, exhibit Na/Cl ratios smaller than 0.60, and the ratio decreases with an increase in chloride concentration (Leonard and Ward, 1962). Brine contamination of fresh ground water in parts of south-central Oklahoma is indicated by chloride concentrations of greater than or equal to 400 mg/L, bromide concentrations greater than or equal to 2 mg/L, the ratio of Na+Cl over TDS is greater than or equal to 0.64, the ratio of Li to Br is less than or equal to 0.01 at Cl concentrations greater than or equal to 400 mg/L, a Na/Cl ratio of approximately 0.46, a Na/Br ratio of approximately 0.92, and a Br/Cl ratio of approximately 0.0048. Chloride and bromide are the most reliable brine indicators. Water-quality degradation in the area may be caused mainly by oil-and gas-field activities (Morton, 1986).

A large area near Crescent was contaminated by oil-field brine which seeped into terrace sands through evaporation pits (Oklahoma Water Resources Board, 1975). It is estimated that possibly more than 100 years will be required to naturally flush out the saline water.

Shallow soil and ground water have been contaminated by seepage from a brine-disposal pit near Burns Flat, Washita County (Shirazi and others, 1976). Inadequately plugged wells polluted ground water near the community of Sasakwa (Smith, 1989).

Pennsylvania: In their report, Miller and others (1974) stated that salt water produced in the state is usually disposed of in unlined surface ponds or is discharged into the ground where it can infiltrate into shallow fresh water aquifers. Several wells and springs in Venango County have been contaminated by this kind of brine discharge.

Oil and gas activities have led to ground-water pollution occurrences that range in aerial extent from 1 acre to more than 50 square miles (U.S. Geological Survey, 1984). Because of the geomorphology of the Appalachian Plateau characterized by gently dipping strata incised by major streams, any contamination occurring in upland areas can cause regional ground-water contamination along the flow paths toward discharge areas (that is, in valleys) (Miller and others, 1974).

Leakage of acid mine drainage through old oil and gas wells and other open holes has caused extensive ground-water pollution in the coal fields of the state (Gass and others, 1977).

South Dakota: In 1950 it was estimated that approximately 12,000 to 15,000 artesian wells within the state leak water into overlying aquifers above them. Inadequately plugged test holes drilled for oil, gas, and uranium may permit upward migration of saline water even in areas where no production is occurring. Salt and gypsum beds in the Spearfish Formation have to be "salted up" during drilling, introducing the potential of salt-water contamination from unlined drill pits. In 1980, there were approximately 27 brine-disposal pits in operation. In 1985, only 50 percent of the produced brine was injected into deep formations (Meyer, 1986).

Leakage from an abandoned well caused salinization of a municipal well in Avon (Jorgensen, 1968, Gass and others, 1977).

Texas: More than 1.5 million wells have been drilled in the state in search for oil and gas (Texas Water Commission, 1989). An inventory of brine production and disposal on the more than 67,000 oil and gas leases throughout the state revealed that a total of 2,237,000,000 barrels of salt water were produced during 1961. Of this total, about 68.7 percent (1,537,000,000 bbls) were re-injected into the subsurface, about 20.6 percent (461,000,000 bbls) were disposed of into open surface pits, 10.1 percent (225,000,000 bbls) were disposed of into surface water, and 0.6 percent (14,000,000 bbls) were disposed of by miscellaneous methods such as spraying on leases and highways (Miller, 1980; Texas Water Commission, 1989). As of 1962, 20,000 to 30,000 injection wells were in operation. In the early years of oil and gas operations, casing programs in hundreds of oil and gas fields were insufficient to protect fresh ground-water resources. Serious regional contamination problems may have occurred in areas where highly pressured brine aquifers occur, such as the Rustler Formation of southwest Texas, the Coleman Junction Limestone in west-central Texas, and the deep Miocene brine aquifers of the Gulf Coast. Plugging of boreholes often was done by simply putting wood, mud, or rocks into the hole and dry holes were often left uncased (McMillion, 1965).

In the early days of exploration, oil-field brines were often diverted into surface streams, as reported from fields in Orange and Hardin Counties. Brines were collected at central points and then pumped through pipelines or allowed to flow through open ditches to estuaries of the Gulf of Mexico (Schmidt and Devine, 1929). In those days, brines were often disposed of onto the land surface by purposely spreading, adding large amounts of salt to soil in oil and gas fields and causing "vegetative kill" areas in many instances. Leaching of salt from such areas may continue to contribute salinity to surface and ground water.

Hundreds of instances of contamination of fresh-water wells as a result of oil-field-brine disposal into surface ponds on the surface of the Ogallala Formation have been reported. Burnitt and others (1963) specify some counties (and instances), such as Andrews County (1 occurrence), Cochran County (6), Ector County (4), Gaines County (11), Garza County (9), Glasscock County (2), Hockley County (9), Hutchinson County (1), Lamb County (1), Lubbock County (3), Lynn County (2), Moore County (1), Terry County (5), and Yoakum County (6). In Winkler County water wells were contaminated by brine disposal into unlined pits and into a communal disposal lake (Garza and Wesselman, 1959). In addition, waste disposal from industrial plants has contaminated wells in Carson, Hockley, Ector, Howard, Moore, Terry, and Yoakum Counties (Burnitt and others, 1963). Disposal of oil-field brine in surface pits has also caused contamination of shallow water wells in Clemville, Matagorda County. Chloride concentrations in affected wells range from 940 ppm to 4,000 ppm (Shamburger, 1958).

Many streams and adjacent ground-water units have been contaminated by salt water as a result of oil-field brine disposal, especially the Brazos and Pecos Rivers. Contamination from brine pits has been reported in Baylor, Cochran, Colorado, Comanche, Cooke, Dawson, Ector, Gaines, Glasscock, Harris, Karnes, Knox, Montague, Pecos, Matagorda, Runnels, Rusk, Victoria, Wilbarger, Wilson, and Winkler Counties, just to name a few. Contamination problems associated with brine-disposal wells are known from Coleman, Karnes, Shackelford, Victoria, and Wilbarger Counties. Improper plugging of test holes may allow downward movement of saline water into the Trinity Sands near the town of Sherman. Unplugged wells in areas underlain by artesian brine aquifers have also resulted in flowing salt water in Knox, Hopkins, and Young Counties, and allowed intrusion of salt water into fresh-water zones in Crockett, Duval, Fisher, Glasscock, Runnels, and Scurry Counties. Contamination by natural gas through faulty well construction has been reported in Caldwell, Bastrop, Comanche, and Wharton Counties, whereas an abandoned well in the Trinity Bay area of Chambers County allowed sea water to contaminate a fresh-water zone (Scalf and others, 1973). In the area of Harrold, brine contamination may have been caused by corroded casing in brine-injection wells, upward flow of injected brine through unplugged, abandoned wells, or disposal of brine into unlined surface pits (Fink, 1965).

Salt-water spills have damaged exceptionally large areas of soil in the following counties (affected acreage in parenthesis): Clay (35,000), Crane (31,250), Ector (15,000), Howard (82,700), Pecos (83,510), Ward (21,500), Winkler (21,940), and Young (18,800) (Texas State Soil and Water Conservation Board, 1984).

Utah: Oil-field brine injection has locally contaminated ground water in the Dakota, Entrada, and Navajo Sandstones of the Montezuma Canyon area (Kimball, 1987). Approximately 97 percent of all brine disposed of into unlined surface impoundments in the Greater Altamont-Bluebell field is lost to seepage into the shallow aquifer system causing local contamination of shallow ground water (Baker and Brendecke, 1983).

West Virginia: Historic records report on salt and oil springs and shallow brine occurrences at various localities in West Virginia. Among those are Campbells Creek, Kanawha County, and Bulltown, Braxton County. Exploration and production of oil and gas has contributed to the salinization of shallow ground water and of surface water. As much as 80 bbls of brine are produced with every barrel of oil; the brine is reinjected into the Injun Sand and the shallower Salt Sands (Bain, 1970).

In the past few decades, salt-water migration toward the land surface has been caused by vertical leakage along hundreds of unplugged wells and test holes. These wells had been drilled during exploration for brine, oil, gas, and coal, and commonly were abandoned uncased or improperly plugged. In Fayette County, chloride concentration of ground water increased from 53 mg/L to greater than 1,900 mg/L within 5 1/2 years due to fresh-water pumpage and inflow of salt water from abandoned boreholes. Secondary recovery of oil in an area of Kanawha County caused accelerated migration of brine upward into fresh-water resources. Oil-field activities also caused an increase in chloride concentration from 100 mg/L to 2,950 mg/L in an aquifer near Wallace, Kanawha County. Hydraulic connection between a brine aquifer and an overlying fresh-water aquifer resulted in upward migration of salt water in oil field areas of Roane County in direct response to pressure-injection disposal of produced brine, causing chloride increases in fresh-water wells from less than 25 mg/L to more than 1,500 mg/L within several months (Wilmoth, 1972).

Problems of oil seepage, old and leaky wells, and brine disposal in the Sisterville, Bens Run, St. Mary's, Belmont, and Waverly oil fields exist between Paden City and Waverly (Carlston and Graeff, 1955). Brine discharge to the river in the Sisterville oil field contaminated the alluvial aquifer, where salt water is reported just above bedrock at 65 to 70 ft. Brine also contaminated the alluvial aquifer through abandoned and leaky wells. Similar conditions were reported at Cox Landing, Ceredo, and Matamoros (Carlston and Graeff, 1955).

Wyoming: Major salt-water problems in the state are associated with oil and gas activities but not with naturally occurring saline ground water (Atkinson and others, 1986).

Mixing of saline water and fresh water due to salt-water intrusion along abandoned oil wells has occurred locally. Degradation of water quality has also been reported to have occurred as a result of irrigation-return flow (Newport, 1977).

Leakage of hydrogen-sulfide gas and salt water into fresh water through oil-well holes has been reported in the Bighorn Basin, especially in an area east and north of Worland. Other areas of oil-field activities are located in the Laramie, Hanna, Wind River, Green River, and Powder Basins (van der Leeden and others, 1975).

3.5 Agricultural Sources

3.5.1 Mechanism

Contamination of surface and ground water from agricultural activities may be associated with irrigation, animal wastes, and commercial chemicals, such as fertilizers, pesticides, and herbicides. These sources may all result in increased salinity of surface and ground water. They may also occur in regions where alternate sources of salinity are present. Because contamination of agricultural sources can result in increased salinities, they need to be identified, especially in the contest of differentiating them from other salinity sources. Agricultural management techniques can also lead to development of saline seep, which will be discussed in chapter 3.6. The following section will not deal with ground-water contamination from agricultural chemicals, such as pesticides, herbicides, and fertilizers, which were discussed in detail by O'Hare and others (1985) and by Canter (1987).

Under natural conditions, a balance exists between the amount of salt entering the soil and the amount of salt that is removed. This balance maintains a certain quantity of salt in the soil that is needed for and tolerated by local vegetation. Change from natural vegetation to agricultural crops and application of irrigation water adds salt to the system. Irrigation can deteriorate ground-water quality in two ways, by inflow of saline water in response to heavy pumpage and by irrigation-return flow. Inflow of saline water as a result of pumpage is included in chapter 3.1 (naturally-saline ground water). The effects of irrigation-return flow on surface- and ground-water quality are discussed below.

Irrigation-return flow water is water that has been diverted for irrigation purposes, was not consumed by processes such as evaporation and transpiration, and finds its way back into surface or ground-water supplies. Irrigation without proper drainage, at insufficient or excessive amounts, the use of poor-quality irrigation water, and solution of surface and soil salts can cause the same effect on soil- and ground-water salinization. Many irrigation projects have caused serious problems of waterlogging and salinity because the relationship between irrigation and drainage had not been realized (Food and Agriculture Organization of the United Nations, 1973). Irrigation-return flow becomes concentrated in chemical constituents from a variety of sources, such as evapotranspiration, solution of minerals, and solution of agricultural residues, such as animal wastes, fertilizers, herbicides, and pesticides. Major processes that occur at land surface are evaporation, solution, and erosion. The degree to which these processes are active depends on factors such as type of soil and rock material, topography, climate, irrigation method and rate, vegetative cover, and the quality of the applied water (Balsters and Anderson, 1979). Infiltrating irrigation water is subject to changes caused by transpiration, evaporation, leaching, ion exchange, and filtration. Approximately 60 to 65 percent of the supplied irrigation water is consumed by growing crops (Law and others, 1970). As water is consumed by plants, most of the salts stay behind. Where natural precipitation is low or drainage is inadequate, these salts accumulate in the soil. This salt has to be removed by an adequate drainage system to permit continued plant growth. Excess irrigation or precipitation water can leach the salt down the soil column to the ground water where it travels to water wells or to natural discharge areas (Fig. 54). In the USSR, five stages of ground-water salinization resulting from irrigation are recognized: (1) an increase in salt concentration during the first years as native soil salts are dissolved; (2) a possible reduction of salt due to higher rates of salt removal than salt dissolution; (3) increase in salinity due to evaporation from a shallow water table which rose to (6–9 ft) below land surface;

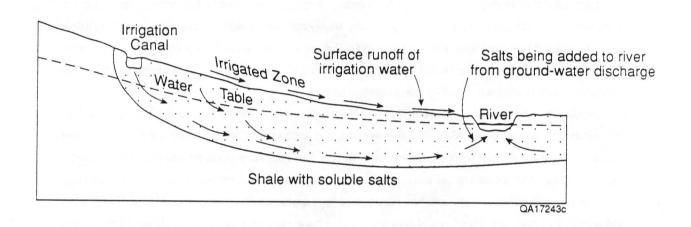

Figure 54. Transport of salt to discharge areas as a result of irrigation-return flow (from Van der Leeden and others, 1975).

(4) reduction through improved artificial drainage; and (5) steady-state conditions of stabilized ground-water salinity (Framji, 1976).

Irrigation with surface water is a major cause of soil and ground-water salinization (Unger, 1977). This problem is especially acute in Montana, Wyoming, Washington, Utah, and North Dakota, where 90 percent of the cropland irrigation water is derived from surface water, as well as in California, Arizona, and New Mexico, where more than 50 percent of the irrigation water is derived from surface-water bodies (Unger, 1977). Salt concentrations in return flows are from two to seven times higher than in the originally applied irrigation water (Utah State University Foundation, 1969). In most instances, these increases are within permissible limits. However, where salt content in the soil is high due to insufficient natural leaching, such as in arid and semiarid areas of the western United States (Table 9), salt content in irrigation-return water can be relatively high. Where drainage is impeded by a high water table, concentrations of soil solutions may be 40 to 80 times higher than concentrations in the irrigation water (Doneen, 1966).

Saline-soil problems may develop in agricultural areas that are low saline and well drained under natural conditions but inadequately drained for additional ground water from irrigation practices. In such a case, the water table might rise within a few feet of land surface, from which water evaporates easily (see also chapter 3.6 on saline seep). This increase in salinity within the shallow subsurface, the root zone, can lead to salinities too high to support plant growth. Similarly, application of high-TDS irrigation water on soils with poor internal drainage can lead to serious soil-salinity problems (U.S. Department of Agriculture, 1983).

Of the total withdrawal of ground water in the United States during 1970, approximately 35 percent was used for irrigation purposes (Geraghty and others, 1973). This percentage is higher in agricultural-dominated states, such as Nebraska, where 85 percent of all ground water pumped in 1970 was used for irrigation (Engberg and Druliner, 1988). Approximately 50 million acres were irrigated nationwide in 1970, with approximately 90 percent of that irrigated area occurring in 17 western states (Fig. 5), where precipitation is insufficient for crop production and where evaporation is high. Irrigation-return flow is of less concern in the eastern states, along the Gulf Coast, and in the Mississippi Valley, where precipitation is plentiful and provides a high degree of dilution and sufficient leaching. Associated with the leaching of soils by excess irrigation water is the flushing of dissolved nitrogen compounds (mostly nitrate) into ground water. Sources of nitrogen are natural soil compounds, which are oxidized to nitrate and dissolved in water preferentially under irrigated and cultivated land, animal wastes and septic tank effluent, and fertilizers (Kreitler and Jones, 1975; Kreitler, 1979).

Pollution of water by animal wastes is increasing because of the increasing number of animals being raised and because of modern methods used in the livestock industry, which may result in higher animal populations. Major sources of pollution are beef cattle, poultry, swine-feeding, and dairy industries (Miller, 1980). Problems associated with these industries include (a) the large supply of nutrients supplied to

Table 9. Percentage of saline and sodic areas in seventeen western states and in Hawaii (from Utah State University Foundation, 1969)[1].

State	Area reported	Total acreage[2]	Saline (all classes) Acres	Percent
Arizona	Statewide	1,565,000	398,830	25.5
California	Statewide	11,500,000	3,744,951	32.6
Colorado	Statewide	2,811,532	981,828	34.9
Hawaii	7 areas	117,418	45,550	38.8
Idaho	All but 3 counties	1,880,063	252,945	13.5
Kansas	Statewide	421,545	102,330	24.3
Montana	4 areas	1,242,728[3]	197,671	15.9
Nebraska	Statewide	1,218,385	290,000	23.8
Nevada	Statewide	1,121,916	475,600	42.4
New Mexico	Statewide	850,000	191,000	22.5
North Dakota	6 areas	2,636,500[3]	816,630	31.0
Oklahoma	Statewide	826,650	193,750	23.4
Oregon	Statewide	1,490,394	103,361	6.9
South Dakota	Statewide	1,697,974	1,196,266	70.5
Texas	4 areas	2,198,950	275,854	12.5
Utah	7 areas	1,390,222	512,782	36.9
Washington	23 counties and the Columbia Basin	2,221,484	266,254	12.0
Wyoming	Statewide	1,261,132	279,703	22.2
TOTAL		36,451,893	10,325,305	28.4

[1]Unpublished data from the U.S. Salinity Laboratory
[2]Irrigable
[3]Arable

aquatic systems and the consumption of dissolved oxygen in aquatic systems, (b) bad odor, and (c) the addition of chemical and biological constituents to water resources (Sweeten, 1990).

The use of chemical fertilizers has doubled from 20 million tons in 1950 to 40 million tons in 1970, with the heaviest use in the Midwest, Texas, and the Sacramento and San Joaquin Valleys of California (Miller, 1980). In terms of tonnage applied, Illinois was the heaviest user, followed by California, Iowa, and Texas. Commercial fertilizer contain nutrients such as nitrate, phosphate, and potassium, which are only partly used by crops. Saffigna and Keeney (1977) reported crop recovery of less than 50 percent of the applied fertilizer. The unused portion finds its way into surface water by runoff from excess precipitation or irrigation and into ground water by recharge to the water table.

3.5.2. Water Chemistry

Constituents commonly analyzed in irrigation water include major cations (Ca, Mg, Na, K), major anions (HCO_3, SO_4, Cl), and the minor constituents NO_3 and B. Total salinity affects the compatibility of the water with the type of crop being irrigated, as certain crops can only tolerate certain levels of salinity. However, in most instances, it is not the absolute salt concentration of the water that is too high, but the accumulation of salt in soil over time, which causes injury to vegetation (Price, 1979). As water salinity increases so, generally, does sodium. The interaction of sodium with clay minerals causes adverse soil changes ("sodium hazard"), that is, exchangeable sodium tends to make a moist soil impermeable to air and water. On drying, this soil is hard and difficult to till. This sodium hazard is expressed as "sodium adsorption ratio," which represents the relative activity of sodium ions in exchange reactions with soil (U.S. Salinity Laboratory, 1954) as determined from the ratio $Na/[(Ca+Mg)/2]^{1/2}$. The relationship between salinity (electric conductivity) and sodium adsorption ratio is commonly used to classify irrigation water in terms of its applicability whereby the hazard increases with increasing salinity and the sodium-adsorption ratio (Fig. 55). A decrease in calcium and magnesium concentration increases the salinity hazard. Calcium may be lost as a result of carbonate precipitation. Magnesium carbonate is more soluble and, therefore, is less likely to precipitate. However, magnesium enters the exchange complex of the soil, replacing calcium (Wilcox and Durum, 1967). This reaction is enhanced by coprecipitation of calcium carbonate. Such a loss of calcium and magnesium causes a relative increase in sodium concentration.

Frequently, boron is analyzed in irrigation water. This is done because of its toxicity to plants even at concentrations of a few mg/L.

Comparison of water-quality measurements between flow into and out of a ground-water basin highlights phenomena taking place within the unsaturated and saturated zones. Irrigation-return waters are typically characterized by net decreases in Ca, HCO_3, and SO_4 as a result of precipitation and by a net gain of sodium, potassium, and chloride (Smith, 1966). In humid areas, the amount of irrigation and dissolved-solids contents are generally low. The proportion of sodium and potassium salts in the drainage

147

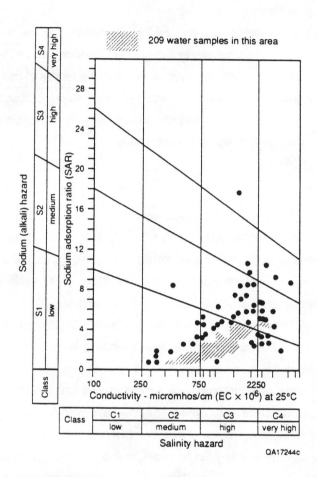

Figure 55. Classification of irrigation water based on sodium-adsorption ratios and conductivity. The higher the conductivity and the sodium-adsorption ratio, the higher the hazard involved (from Price, 1979).

148

water is small when compared to iron and aluminum. In contrast, in arid areas soils contain a considerable amount of leachable Na, K, Ca, and Mg salts (Utah State University Foundation, 1969).

In some areas, low-chloride brines from oil-and gas-field areas are used for irrigation. These brines need to be analyzed for constituents other than just major cations and anions because of the often high content of trace metals in oil-field waters. The high bicarbonate content in some low-chloride waters produced with methane gas (for example, San Juan Basin, New Mexico) can increase the sodium hazard because these waters are typically very high in sodium and very low in calcium and magnesium concentrations.

Soil salinity is commonly expressed in conductivity terms as micromhos/cm. Conductivity ranges and corresponding salt contents are listed in Table 10. Salinity expressed in units of microsiemens/cm can be converted to ppm or mg/L by multiplying by a factor of 0.6.

3.5.3. Examples of Geochemical Studies of Agricultural Salinization

Irrigation is practiced along the Arkansas River in Colorado. As water use and re-use increases downstream, so does the concentration of TDS (Hearne and others, 1988) (Fig. 56). This increase may change the chemical composition of individual constituents to a degree that makes it easily distinguishable from background water quality using Stiff diagrams (Fig. 57).

Irrigated agriculture is responsible for chloride and nitrate concentrations above natural background levels in parts of central Wisconsin (Saffigna and Keeney, 1977). Excess fertilizer is leached by excess irrigation water because crops recover only 50 percent of the fertilizer. Absolute concentrations of chloride and nitrate vary to a high degree in direct relation with the time of fertilization and of irrigation application. Ratios of Cl/NO_3, in contrast, are relatively uniform, suggesting a common source. In fact, applications of chloride-rich fertilizer are similar to application rates of nitrogen fertilizer. Because some of the nitrogen is consumed by plants, whereas chloride is not, ratios of Cl/NO_3 are generally greater than one. Potassium, another element enriched in the fertilizer (KCl), does not reflect this source of contamination because of its uptake by plants (Saffigna and Keeney, 1977).

The storage and disposal of agricultural chemicals may be more critical than the application of these chemicals on fields (Waller and Howie, 1988). In Dade County, Florida, after 50 years of agricultural activity, fertilizer application has increased nitrate and potassium concentrations in ground water to levels that are usually below health standards. But in areas of storage and dumping of agricultural chemicals, ammonia, potassium, and organic nitrogen were one order of magnitude and nitrate was two orders of magnitude above background levels. These parameters allow distinction from other salinization sources in the area, that is, inflow of artesian saline water. Elevated concentrations of magnesium, sodium, and chloride are associated primarily with this second source. Elevated in disposal areas were also iron and manganese,

149

Table 10. Soil salinity classification in the United States (from Food and Agriculture organization of the United Nations, 1973).

Classification	Conductivity	Total Salt Content
Salt free	0–4 millimhos/cm	0–1,500 mg/L
Slightly saline	4–8	1,500–3,500
Moderately saline	8–15	3,500–6,500
Strongly saline	>15	>6,500

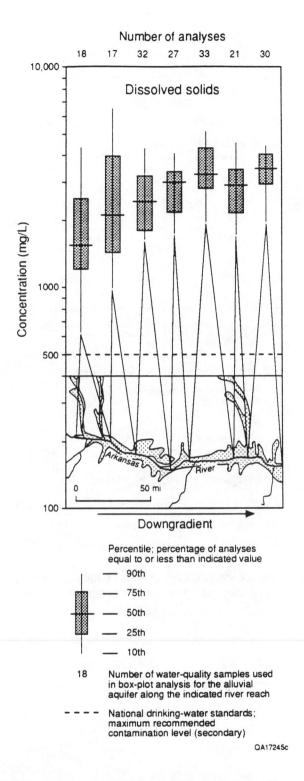

Figure 56. Increase in TDS concentration in the Arkansas River alluvial aquifer as a result of irrigation-return flow (from Hearne and others, 1988).

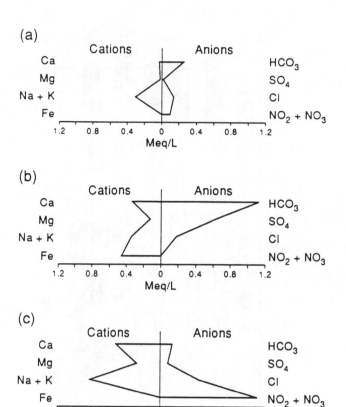

Figure 57. Use of Stiff diagrams for identification of water-quality changes as a result of agricultural activities: (a) (b) background water quality; (c) water affected by agriculture (from Denver, 1988).

152

whereas other trace elements, such as arsenic, cadmium, chromium, copper, lead, mercury, nickel, and zinc occurred below detection limits (Waller and Howie, 1988).

The impact of irrigation-return flow on surface-and ground-water quality were investigated experimentally by Law and others (1970). Surface-return flow after three irrigation periods showed increases in TDS of between 8.5 and 32.2 percent above irrigation-water concentrations (Fig. 58a), with the degree of change being related to weather conditions between irrigation applications. The large change between the June 28 and the July 18 irrigations (Fig. 58a) was caused by a dry period during which salt accumulated in the soil. The salt was dissolved and carried away with the July 18 irrigation. A rain event of 1.6 inches between the July 18 and the August 18 irrigations, in contrast, removed soil salts, causing only a slight TDS increase in runoff after the third irrigation (Fig. 58a). This alternating change in water quality is not reflected in water percolating through the soil, as indicated by a gradual TDS increase as the irrigation season progresses (Fig. 58b).

Increases in soil salinity at this site are many times higher, up to twentyfold at a depth of 18 to 24 inches, than increases under nonirrigated land (Fig. 59a). These increases are associated with downward displacement of salts, as shown by the relationship of sodium and chloride with depth (Fig. 59b). The magnitude of difference between the nonirrigated and the irrigated profiles was caused by salts brought in by the irrigation and by evapotranspiration. Sulfate, calcium, and magnesium concentrations don't follow the simple depth relationship seen for sodium and chloride, but instead exhibit three significant peaks (Fig. 59c). Natural occurrences and the solution of gypsum in the soil may control this distribution with depth.

The chemical composition of irrigation-return water reflects local conditions and the irrigation history of an area. For example, in the Yakima Valley of Washington, sodium and chloride concentrations show higher net increases between applied irrigation water and outflow drainage (Na_{in}/Na_{out} = 4.4; Cl_{in}/Cl_{out} = 6.5) than magnesium and sulfate concentrations (Mg_{in}/Mg_{out} = 2.0; SO_{4in}/SO_{4out} = 3.6) (Sylvester and Seabloom, 1963). The overall increase reflects drainage of an area formerly waterlogged and saline because of previous irrigation practices (Utah State University Foundation, 1969). In the Grand Valley of Colorado, in contrast, net changes between inflow and outflow show large increases in magnesium (Mg_{in}/Mg_{out} = 11.4) and sulfate (SO_{4in}/SO_{4out} = 10.8), a relatively small increase in sodium (Na_{in}/Na_{out} = 2.5), and even a decrease in chloride loads (Cl_{in}/Cl_{out} = 0.6). This net change is due to replacement of formation water in salt-rich shales and to reactions between irrigation water and soil minerals (Skogerboe and Walker, 1973).

Irrigation and petroleum production account for ground-water deterioration in the High Plains Aquifer of south-central Kansas. Irrigation-affected ground water is characterized by increased concentrations of calcium, magnesium, potassium, fluoride, and nitrate, whereas ground water affected by oil-field brine is characterized by increased concentrations of TDS, sodium, and chloride (Helgeson, 1990).

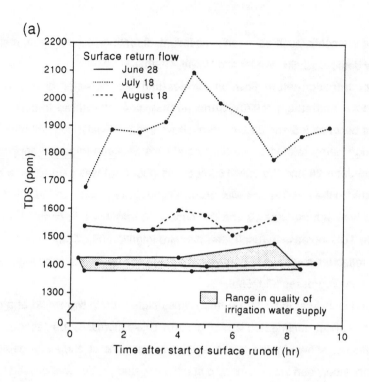

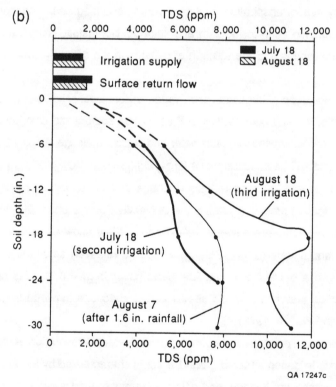

Figure 58. Salt content in (a) irrigation water and irrigation runoff and in (b) soil during and after three controlled irrigation experiments (from Law and others, 1970).

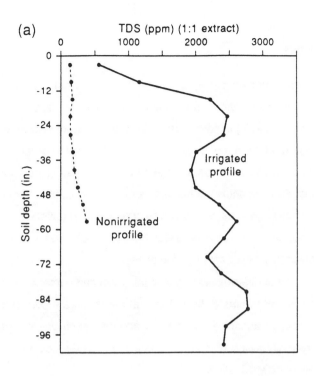

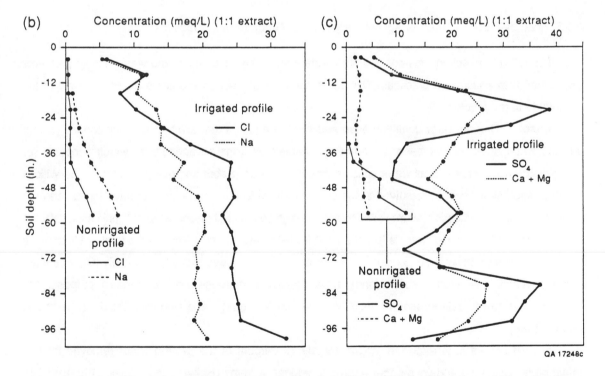

QA 17248c

Figure 59. Relationship between (a) TDS, (b) Na, Cl, and (c) Ca+Mg, SO$_4$ and depth in soil under irrigated and nonirrigated land. Downward displacement of salts occurs under both profiles, with differences in magnitude caused by salts brought in by irrigation, by evapotranspiration, and by the natural occurrence of gypsum in the soil (from Law and others, 1970).

155

3.5.4. Significant Parameters

Degradation of ground-water quality by agricultural activities can be caused by (a) solution and transport of chemicals, such as herbicides, pesticides, and fertilizers, (b) disposal of animal waste and waste water from animal farms, and (c) irrigation-return flow. With respect to ground-water salinity, irrigation-return flow is the most important source of degradation. Evapotranspiration and leaching of soil minerals accounts for increases in most chemical components in drainage waters from irrigated areas. Typically, chloride and sodium concentrations show the highest increases, although other constituents may be high in some areas, reflecting local conditions. Significant parameters in irrigation-return flow may change over time, as original soil minerals are dissolved in the initial irrigation stage of an area and minerals brought in by irrigation water are dissolved in subsequent irrigation phases.

A significant parameter that differentiates agriculturally induced contamination from other salinization sources discussed in this report is nitrate. In agricultural areas, nitrate concentrations are often above background values. Salinization associated with other sources, such as sea-water intrusion or oil-field pollution, in contrast, is typically associated with increases in chloride, sodium, calcium, and magnesium concentrations and with small NO_3/Cl ratios.

3.5.5. State-by-State Summary of Agriculturally Induced Ground-Water Problems

The following section provides a state-by-state summary of some of the water-quality problems associated with agricultural practices. This list is not complete but should serve as an overview of the magnitude of the problem.

Arizona: Ground-water quality in deep and shallow aquifers that are hydraulically connected with surface water supplies has been deleteriously affected by irrigated agriculture, which accounts for approximately 90 percent of all water consumption in the state (Sabol and others, 1987). In the Willcox Basin of southeast Arizona, dissolved-solids concentrations in the alluvial aquifer have increased as a result of recharge of irrigation-return water containing high contents of dissolved salt. This is especially the case where depth to ground water is less than 100 ft (Kister and others, 1966). Of the total salt load in the Colorado River at Hoover Dam (9 million tons annually), it is estimated that approximately 37 percent are contributed by irrigation-return water (Jonez, 1984). Recycling of ground water for irrigation purposes has caused water-quality deterioration in the Wellton-Mohawk Irrigation and Drainage District (Effertz and others, 1984).

Arkansas: As a general rule, approximately 75 percent of the ground water removed for rice irrigation is consumed and 25 percent returns to aquifer systems (Holland and Ludwig, 1981). Water-quality deterioration from irrigation-return water has occurred along the Arkansas River (Scalf and others, 1973).

156

California: Agriculture is extensive throughout most of the Central Valley, parts of Imperial, Riverside, and San Bernardino Counties, and many coastal and Southern California counties (Lamb and Woodard, 1988). Ground-water contamination by pesticides is widespread throughout these areas, as indicated by a monitoring survey of water wells from 1979 through 1984. Of 8,190 wells tested, 2,522 (31 percent) were contaminated by pesticides (Lamb and Woodard, 1988). In the San Joaquin Valley, selenium concentrations in agricultural drainage waters are high, exceeding 1 mg/L in some places (Deverel and Gallanthine, 1989).

Approximately 32 million gallons of fresh water are used daily for irrigation and livestock in California (Miller, 1980). Irrigation water contributes an average of approximately 1.18 tons of salt per acre per year (Thorne and Peterson, 1967). Degradation of ground water as a result of irrigation has been reported from the San Joaquin Valley (Fuhriman and Barton, 1971), where imported irrigation water adds nearly 2 million tons of salt to soil and water each year (U.S. Geological Survey, 1984). Irrigation-return water also causes TDS increases in some of the southern basins, such as in the Ventura Basin (TDS as much as 2,500 mg/L) and the Salinas Basin (TDS as much as 1,000 mg/L) (U.S. Geological Survey, 1984).

Colorado: Irrigation-return waters have caused TDS increases in ground water in the High Plains aquifer, the San Luis Valley aquifer system, the South Platte alluvial aquifer, and the Arkansas River alluvial aquifer (Hearne and others, 1988). Generally, shallow ground water appears to be more affected by irrigation-return water than deep ground water. In parts of the San Luis Valley, insufficient drainage of saline soils had led to land abandonment in the early 1900's (Siebenthal, 1910). Approximately 37 percent (700,000 to 800,000 tons per year) of the total salt load in the Upper Colorado River Basin is attributed to saline flow from irrigation in the Grand Valley of western Colorado. This contribution is a result of excess irrigation water and seepage from irrigation canals dissolving salts and subsequently discharging into the river (Skogerboe and Law, 1971; van der Leeden, 1975). The repeated use of surface water (as much as seven times) for irrigation purposes within a 65-mi stretch from Denver to Kuner caused an increase in mineralization of surface and ground water in the South Platte River Valley (van der Leeden and others, 1975).

Irrigation and feedlots contribute to ground-water contamination in Weld County (Rold, 1971). High nitrate concentrations from animal waste or fertilizer leachate have been reported along Black Squirrel Creek, the High Plains aquifer (Hearne and others, 1988), and in the San Luis Valley aquifer (Edelmann and Buckles, 1984).

Delaware: Coastal Plain aquifers have been affected by agricultural nutrients, such as nitrate and nitrite, especially downgradient from poultry farms (Ritter and Chirnside, 1982). Most of the known problem areas are located in Sussex and Kent Counties (Denver, 1988). Other constituents above background levels as a result of agricultural activities include chloride, calcium, sodium, and potassium (Denver, 1988).

Florida: Of all the eastern states, Florida uses by far the most water for irrigation. This is because the state's rainfall is concentrated within a few months of the year, whereas the climate permits the growing of crops more or less year-round (Geraghty and others, 1973). Associated with long growing seasons is the heavy use of pesticides, which has affected more than 1,000 public and private water-supply wells (Irwin and Bonds, 1988).

In Dade County, storage and dumping of agricultural chemicals contributes more severely to high concentrations of ammonia, potassium, and organic nitrogen in ground water than does the application of these chemicals onto fields (Waller and Howie, 1988).

Illinois: It is estimated that 97 percent of all rural-domestic water systems are supplied from shallow aquifers (Voelker and Clarke, 1988). These shallow aquifers are especially vulnerable to contamination by irrigation-return waters.

Kansas: Irrigation-return flows have caused increased concentrations of calcium, sodium, sulfate, and chloride in ground water of north-central Kansas ground water (Spruill, 1985).

Montana: Agricultural practices contribute to the widespread occurrence of dryland saline seep throughout most of the state (see also chapter 3.6). Irrigation-return flow with sometimes high concentrations of nitrate have contaminated water wells in Rosebud County (van der Leeden and others, 1975).

Nebraska: Increased usage of irrigation water may lead to future problems in the Big Blue and Little Blue River Basins, the western parts of the Republican River basin, and most of Box Butte and Holt Counties (Exner and Spalding, 1979).

Nevada: Contamination potentials from agricultural sources in Nevada are caused by mineralized irrigation-return water, feedlot and dairy-farm effluents, and agricultural chemicals (Thomas and Hoffman, 1988). Large arsenic concentrations in ground water in parts of Churchill County have been traced to irrigation-induced leaching of soils (Thomas and Hoffman, 1988).

New Mexico: Ground-water problems in deep and shallow aquifers that are hydraulically connected with surface-water supplies have been deleteriously affected by irrigated agriculture, which accounts for approximately 90 percent of all water consumed in the state (Sabol and others, 1987). In the Rio Grande Basin, TDS content increases progressively in a downstream direction, with the concentration at Fort Quitman (1,770 ppm), Texas, being nearly 10 times the concentration at the Otowi Bridge (180 ppm), just below the Colorado–New Mexico state line (Fireman and Hayward, 1955). Water-quality deterioration from irrigation-return water has also occurred along the Pecos River (Scalf and others, 1973).

Oklahoma: Investigations of irrigation-return waters identified several instances of surface and ground-water deterioration. Compared with the quality of water applied, TDS in irrigation-water return flow increased by about 20 percent and in percolating soil water by more than 500 percent (Law and others 1970; van der Leeden, 1975). Water-quality deterioration from irrigation-return water has occurred along the Arkansas River (Scalf and others, 1973).

Texas: In the Rio Grande Basin, TDS content increases progressively in a downstream direction as a result of irrigation-return flows. TDS concentration at Fort Quitman, Texas, (1,770 ppm) is nearly 10 times the concentration at the Otowi Bridge (180 ppm) just below the Colorado–New Mexico state line (Fireman and Hayward, 1955). Water-quality deterioration from irrigation-return water has also occurred along the Pecos River (Scalf and others, 1973) where some wells in the principal irrigation areas of Reeves County have experienced significant increases in TDS (Ashworth, 1990).

Approximately one million acres have been affected by irrigation to the degree that they are considered saline (EC>4 millimhos/cm). Counties with more than 10,000 acres of saline soil are (acreage in parenthesis): Cameron (275,000), Chambers (20,000), Culberson (44,000), El Paso (57,000), Hidalgo (350,000), Hudspeth (68,000), Jefferson (10,250), Maverick (10,000), Reeves (10,000), San Patricio (10,000), Ward (30,000), Willacy (10,000), and Zavala (50,000) (Texas State Soil and Water Conservation Board, 1984).

Utah: Irrigation-return water has caused increased salinity of surface water in the Uinta River Basin (Brown, 1984). In the Price and San Rafael River Basins of east-central Utah, irrigation-return flow led to TDS increases from background levels of 400 to 700 mg/L to contamination levels of 2,000 to 4,000 mg/L (Johnson and Riley, 1984). The same process caused a TDS increase of 2,000 percent in the Sevier River within a 200-mile-long stretch (Thorne and Peterson, 1967). Other areas where ground-water deterioration has occurred in response to irrigation pumpage include the Pahvant Valley and the Beryl-Enterprise area (Waddell and Maxell, 1988), as well as the Curlew Valley, where high pumpage rates from irrigation wells resulted in upward movement of saline ground water from deep aquifers (Bolke and Price, 1969).

Washington: Irrigation-return flows with sometimes high concentrations of nitrate have contaminated water wells in the Odessa area and in Snohomish County (van der Leeden and others, 1975).

Wyoming: Mineralized irrigation-return flows have contaminated ground water in alluvial aquifers along the Shoshone, Bighorn, and Big Sandy Rivers (Wyoming Department of Environmental Quality, 1986).

3.6 Saline Seep

Salinization associated with saline seep is considered separately from the previous section on agricultural sources because of its significance in the Plains Region of the United States and Canada.

3.6.1. Mechanism

Saline seep was defined by Bahls and Miller (1975) as "recently developed saline soils in nonirrigated areas that are wet some or all of the time, often with white salt crusts, and where crop or grass

159

production is reduced or eliminated." It differs from salinization as a result of irrigation-return flow through the dominant salinization mechanism, that is, evaporation from a shallow water table accounts for high salinities in saline-seep systems, whereas leaching of soil salts is the major mechanism that accounts for high salinities in irrigation-return flows.

Intensive studies of the saline-seep phenomenon in the United States have been more or less restricted to the state of Montana, even though the problem exists elsewhere. In many localities in the Northern Great Plains, saline seep causes loss of productive farmland, impassable roads, flooded basements, stock deaths, and salinization of surface and ground water (Custer, 1979). Conditions that favor saline-seep development are widespread across the Northern Great Plains (Fig. 6), covering an area exceeding 200,000 mi^2 throughout most of North Dakota, approximately half the area of Montana and of South Dakota, and several Canadian provinces (Bahls and Miller, 1975).

Several conditions have to be met for saline seep to develop, such as excess percolation of recharge water, soluble soil or aquifer minerals, a low-permeable unit at relatively shallow depths, an internally drained flow system, and evaporation. Agricultural activities often play a significant role in creating or intensifying some of these conditions. Excess percolation is often caused by destruction of natural vegetation and drainage ways (for example, by terracing of land), by excessive irrigation, or by practices such as fallow cropping or by planting of crops that use little water. Undisturbed ecosystems are characterized by optimal environmental adaptation of plant covers that withstand a wide spectrum of environmental changes, such as dry and wet periods or hot and cold periods. A great diversity of natural plant cover provides a certain degree of protection to individual plants. This adaptation and protection is mostly lost in monoculture systems. With regard to saline-seep development, the most important characteristic of undisturbed systems is the efficient uptake of water through the variety of natural plants, which prevents large increases in water-table elevation. Disturbed systems are often less efficient in uptake and allow a greater portion of water to percolate through the soil and to recharge the local ground-water flow system (Bahls and Miller, 1975). Initial stages of saline seep are often indicated by prolonged soil-surface wetness following heavy rains. Fallow areas can undergo a water-table rise of 1 to 15 ft during years of above average spring precipitation (Miller and others, 1980). Although some of this water will be used up during the growing season, low stands of previous years are normally not reached, causing steady increases in water levels over the previous years. In eastern Montana, the water table within the glacial till has been rising an average of four to ten inches per year (Bahls and Miller, 1975). Where the water table intersects or is close to land surface in response to these water-level increases, water-logging or seepage will occur (Fig. 60). A water table within three feet of land surface signals potential problems, as water will move up from this depth to the surface by capillarity (U.S. Department of Agriculture, 1983).

Depending on the mineralogy of material encountered along the flow path, water-logged or seepage areas will be more or less saline. Minerals that are critical in salinization by excess percolating water and by the increased water table include pyrite, sodium-rich clays, carbonates, gypsum, sodium and magnesium

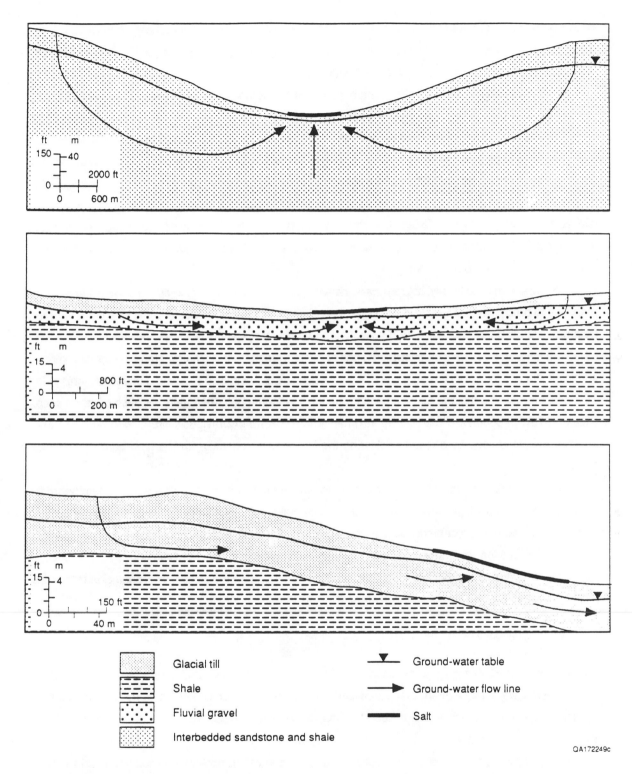

	Glacial till		▼	Ground-water table
	Shale		→	Ground-water flow line
	Fluvial gravel		▬	Salt
	Interbedded sandstone and shale			

QA172249c

Figure 60. Diagrammatic cross section of ground-water flow with saline seep in topographically low areas and in intermediate topographic position (from Thompson and Custer, 1976).

161

sulfate, and nitrate (Thompson and Custer, 1976; Kreitler, 1979). These minerals are normally bound to the soil but are dissolved and transported to discharge areas by excess recharge. Evapotranspiration increases the salinity of discharge water in seepage areas.

An impermeable layer at shallow depths increases the potential of saline seep by preventing deep percolation and by creating a perched water table. The presence of a perched water table is indicated during drilling when a saturated zone is penetrated before passing through powder-dry shale. In Montana, this shale layer is generally encountered within 70 ft of land surface (Thompson and Custer, 1976). Where the shale layer occurs within 30 ft of land surface, the potential for saline-seep development is very high. Where the overlying glacial till is thicker than 30 ft, water-table elevation has not yet reached land surface (Bahls and Miller, 1975). In Texas, a correlation exists between soil texture and seep occurrence. Approximately 80 percent of all saline seeps in that state are controlled by low permeabilities of fine and fine-loamy soils (Neffendorf, 1978).

Seep water may be locally, not regionally, derived, as indicated by the rapid rise in static water level in response to precipitation. The size of the seep (discharge) area is directly related to the size of the upland (recharge) area (Bahls and Miller, 1975). Recharge and discharge areas may be relatively close to each other and small in size or encompass thousands of acres. Runoff from discharge areas can greatly impair surface-water quality. For example, some of the largest rivers of Australia have seen a threefold increase in TDS concentrations within the 50-year period from the 1910's through the 1960's because of saline seep (Peck, 1978). In southern Australia, approximately 430,000 ha of previously productive farmland are affected by saline seep as a result of clearing the indigenous vegetation for farming purposes (Peck, 1978).

In some instances, saline seep is mistaken for oil-field pollution or for saline water emerging from great depths along geologic structures. Characteristics that positively identify saline seep include during drilling, wet material is encountered above a dry substrate, the static water level in saline seeps responds rapidly to precipitation, the water table reflects topography, ground-water chemistry is typical for shallow, local ground water rather than deep ground water, and changes in drainage or cropping practices in recharge areas affect the size of seepage areas (Custer, 1979).

3.6.2. Water Chemistry

During a field check of seep salinities, Custer (1979) noticed that seeps with low specific conductivities were most often associated with sandstone units or colluvium derived from nearby sandstones, whereas seeps with high specific conductance were associated most often with shale units. The increased salt content in the shale units (up to 60,000 micromhos at 25°C) is derived from the weathering of pyrite, solution of carbonates and sulfates, and cation exchange (Thompson and Custer, 1976). Oxidation and dissolution of pyrite creates acidity that becomes available for hydrolysis reactions and solution of carbonates (Donovan and others, 1981) and often leads to high selenium concentrations in ground water. Breakdown of chlorite, illite, and feldspars contributes Mg, K, Na, and SiO_2 to seep waters, which makes for a strong correlation between these constituents and TDS. Sodium is also derived from adsorbed positions in smectites through exchange reactions involving hydrogen and calcium ions (Donovan and others, 1981). Gypsum dissolution may be one of the most dominant leaching processes in water with good correlation between SO_4 and TDS (Fig. 61). These chemical reactions produce a remarkably uniform water chemistry in Montana, from low-TDS (1,500 to 3,000 mg/L), Ca-HCO_3 type waters in recharge areas to high-TDS (4,000 to 60,000 mg/L), Na-Mg-SO_4 type water in discharge areas (Miller and others, 1980). Sulfate concentrations of up to 33,000 mg/L have been reported from test holes in seep areas in the Fort Benton area, Montana (Bahls and Miller, 1975). With the exception of recharge waters, salinization leads to supersaturation with respect to calcite, dolomite, and gypsum in many seep-water systems (Fig. 62). Chloride concentrations are normally relatively low but minor and trace constituents, such as NO_3, Al, Fe, Mn, Sr, Pb, Co, Zn, Ni, Cr, Cd, Li, and Ag, are relatively high, which distinguishes them from most other naturally saline ground waters (Bahls and Miller, 1975).

Saline-seep waters in parts of northeastern North Dakota differ from most seep waters in Montana by a dominance in Mg-Cl type as opposed to Na-SO_4 type (Sandoval and Benz, 1966). A general absence of gypsum accounts for the low sulfate concentrations in those waters, which reach TDS concentrations of up to 34,000 ppm. Water-table elevations in salt-affected areas range from 1.5 to 5 ft (Sandoval and Benz, 1966). Waters associated with saline seep in Australia are generally of the Na-Cl type, in contrast to dryland saline seep in North America (Peck, 1978). Oceanic salts found in rainfall may be an important source of these solutes (Hingston and Gailitis, 1976; Peck, 1978).

White salt crusts frequently occur on the ground at seep areas. Thompson and Custer (1976) identified the minerals Loeweite ($Na_{12}Mg_7(SO_4)_{13}$) and, at several locations, Thenardite (Na_2SO_4). Major chemical processes operating in saline seep systems and discussed above are summarized in figure 63.

3.6.3. Examples of Geochemical Studies of Saline Seep

Thompson and Custer (1976) used triangular diagrams (modified Piper diagrams) for displaying chemical characteristics of seep waters in Montana. These waters are typically of the Na-Mg-SO_4 type, with some Ca-HCO_3 waters (Fig. 64). Ground-water flow from recharge areas to discharge areas induces chemical changes, such as ion exchange, involving Ca in the water and Na on soil particles. Ground-water below dryland-farm sites near Rapelje, Stillwater County, is of the Na-Mg-Ca-SO_4 type, with TDS concentrations of about 9,000 ppm (Thompson and Custer, 1976). Rock weathering is the source of dissolved constituents, including processes like leaking of formation water rich in sodium chloride, oxidation of pyrite, solution of carbonate minerals, exchange of calcium for sodium on exchange sites, and

163

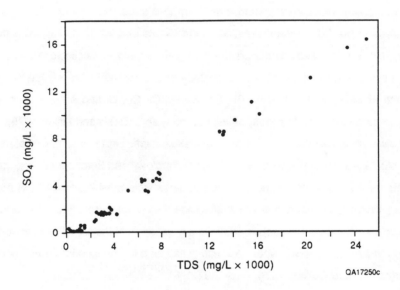

Figure 61. Correlation between SO_4 and TDS concentrations in seep waters from the Colorado Group, Montana, suggesting gypsum solution as the major contributor to salinity (from Donovan and others, 1981).

164

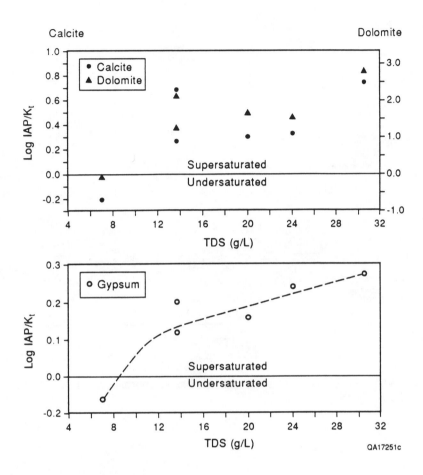

Figure 62. Relationship between saturation states and salinity in well waters affected by saline seep, north-central Montana. Seep salinization typically leads to saturation with respect to calcite, dolomite, and gypsum (from Donovan and others, 1981).

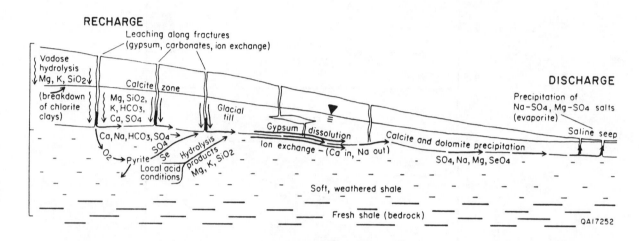

Figure 63. Summary of chemical and transport processes operating in a saline-seep system (from Donovan and others, 1981).

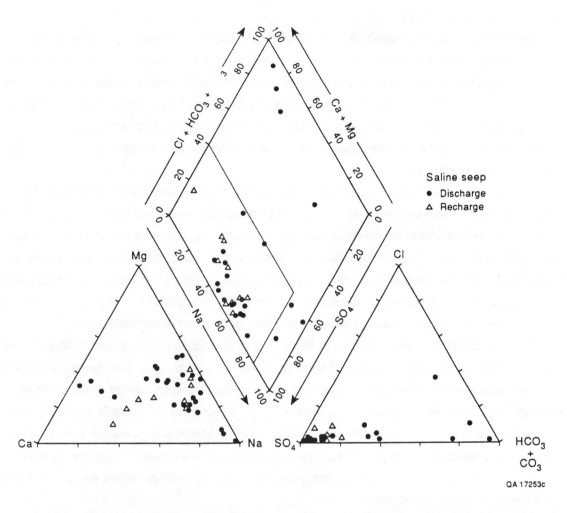

Figure 64. Modified Piper diagram of chemical composition of ground water in saline seeps, Montana. Seep water typically is of a Na-Mg-SO$_4$ type as a result of sulfate-salt solution and ion exchange (from Thompson and Custer, 1976).

167

precipitation of calcium, sodium, and magnesium sulfate. Below cultivated land, high nitrate concentrations have been measured. Deep percolation of recharge water during fallow years leaches the nitrate into ground water. Below uncultivated land, in contrast, nitrate is absent in ground water (Thompson and Custer, 1976).

Experimental leachates of cored material from seepage areas bear little similarity to actual saline-seep areas (Donovan and others, 1981). Apparently, the chemistry of ground water is affected by underlying bedrock or aquifer geology, which changes the composition to a degree that cannot be duplicated in the laboratory. Gypsum dissolution and ion exchange are dominant processes governing the major constituents (Na, Mg, SO_4) in some waters. High bicarbonate concentrations and roughly equimolar Ca/Mg ratios indicate that carbonate dissolution is additionally an important process in other waters (Donovan and others, 1981).

High water tables and poor drainage conditions contribute to high salinity in soils along the Red River of North Dakota. Ground-water salinity of up to 40,000 ppm has been measured in some areas (Benz and others, 1961). The water is of the Mg-Ca-Cl-SO_4 type in the most saline areas and of the Mg-Ca-SO_4-Cl type in moderately saline areas. In contrast to seep areas in Montana, this occurrence of saline water is part of a regional flow system instead of a local flow system; not weathering, but vertical upward flow of saline water from the Dakota sandstone, is the source of salinity. This is indicated by the chemical similarity between Dakota Sandstone Formation water and seep waters (Benz and others, 1961).

The condition of dryland saline seep in Texas is often blamed on oil-field pollution. Although the visual appearances of brine- and seep-affected soils are very similar, both often being void of any vegetation and covered with white crusts, chemical characteristics are quite different. In most instances, total salt content as well as sodium and chloride concentrations are appreciably higher in brine-contaminated areas than in dryland seeps. The difference in chemical composition between seep-affected and unaffected soils in the Central Rolling Red Plains of Texas was investigated by the U.S. Department of Agriculture (1983). All major ions increase significantly in the soil within the upper few feet below seeps, reflecting evaporation.

A variety of salinization sources, including saline seep, natural discharge of saline formation water, and oil-field pollution, is known to occur in parts of West Texas. In a two-county salinization study, Richter and others (1990) developed methods to distinguish saline-seep waters from the other salinization sources. In both counties, fresh ground water (Cl <250 mg/L) is of similar facies type, dominated by Ca, Mg, HCO_3, and SO_4 (Fig. 65). At chloride concentrations greater than 250 mg/L, however, chemical facies differ in those two counties, suggesting that different salinization mechanisms are dominant. While ground water in Runnels County remains a Ca-Mg to mixed-cation water, ground water in Tom Green County exhibits a trend of increasing Na proportions with increasing chloride proportions. The trend toward Na-dominated water may indicate mixing with a Na-Cl-dominated saline water, whereas the unchanged cationic composition may suggest a mechanism that doesn't change ionic proportions, such

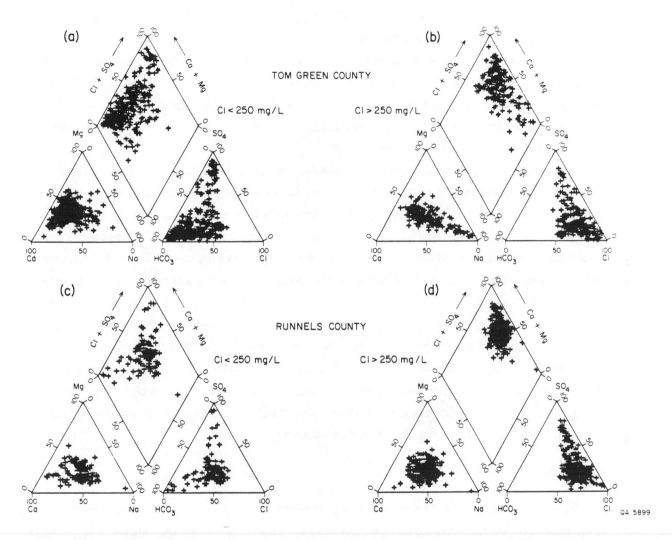

Figure 65. Piper diagram of shallow ground water in two adjacent counties of West Texas. Different salinization mechanisms are suggested from data distributions in the cation triangles. At chloride concentrations greater than 250 mg/L, a mixing trend between fresh Ca-HCO₃ and saline Na-Cl water appears to cause the shift in cation percentage in Tom Green County (a) (b). In Runnels County, in contrast, cation percentages do not change, suggesting a salinization mechanism that doesn't change relative concentration ratios, such as evaporation (c) (d) (from Richter and others, 1990).

169

as evaporation. Testhole samples from the two counties plotted on bivariate plots can be used to identify the possible sources of salinity (Fig. 66). Among samples from Runnels County, slopes of Mg and SO_4 plotted against Cl on logarithmic scales are not significantly different from 1.0 (α = 0.05), whereas among Tom Green County samples, slopes of Ca, Mg, and SO_4 plotted against Cl are significantly different from 1.0 (α = 0.05). The differences are interesting because ground-water evaporation without mineral precipitation gives rise to ionic relationships with unit slope because ionic concentrations increase without changing molar ratios of chemical constituents. Therefore, it is reasonable to suspect that Runnels County samples are influenced by evaporation. Tom Green County samples and subsurface brines have similar ionic relationships, suggesting that the mixing of fresh ground water and brines is an important salinization mechanism in that county (Richter and others, 1990).

In some areas, agriculturally induced saline seep can be distinguished from other salinization sources by consideration of contaminants that are typical for agricultural pollution, such as nitrate. Kreitler and Jones (1975) identified extremely high nitrate concentrations averaging 250 mg/L (range from <1 mg/L to >3,000 mg/L) in shallow ground water of Runnels County, Texas. Nitrogen isotopes of soil and ground-water samples identified natural soil nitrogen and animal waste as the major sources of high nitrate levels. The process that produced these high concentrations in ground water was summarized by Kreitler and Jones (1975) as follows: "Dryland farming since 1900 has caused the oxidation of organic nitrogen in the soil to nitrate. During the period 1900 to 1950, nitrate was leached below the root zone but not to the water table. Extensive terracing after the drought in the early 1950's has raised the water table approximately 20 ft and has leached the nitrate into ground water. Tritium dates indicate that the ground water is less than 20 years old." High nitrate concentrations and the nitrogen-isotopic composition of nitrate indicate that dryland saline seep is the dominant source of salinization in that area and not oil-field brine, which typically is very low (less than 1 mg/L) in nitrate content.

3.6.4. Significant Parameters

Saline-seep water chemistry is governed by evaporation, resulting in an increase of all constituents in the water. The increase is reflected on constituent plots as evaporation trends, in contrast to mixing trends toward a saline-water source observed for the other salinization sources discussed in this report, with the exception of irrigation-return waters. At low salinities, this increase is characterized by more or less constant constituent ratios of major ions, such as Ca/Cl, Mg/Cl, or SO_4/Cl. With increasing salinity, mineral precipitation will change these ratios as carbonates and sulfates begin to form. Precipitation products will vary from area to area depending on the chemical composition of soil and water. Where sources of sulfate are abundant, dissolved sulfate concentrations may by far exceed the concentration of dissolved chloride, which distinguishes seep water from most other saline ground water. Miscellaneous trace constituents

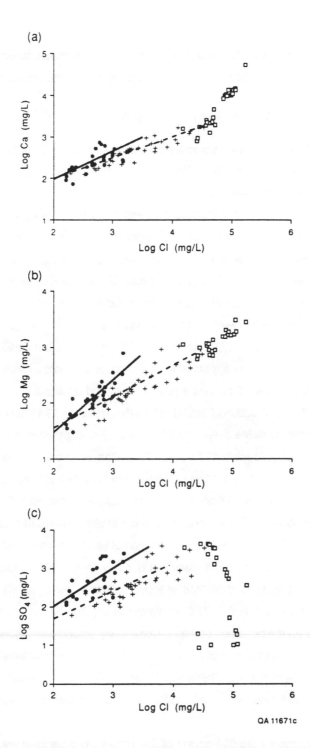

Figure 66. Variation in (a) calcium, (b) magnesium, and (c) sulfate concentrations with chloride concentration in shallow ground water from a two-county area in parts of West Texas. Samples collected from the eastern county (solid dots, Runnels County in Figure 65) plot similar to the theoretical evaporation line of these constituents (unit slope), whereas samples collected from the western county (crosses, Tom Green County in Figure 65) trend toward the composition of subsurface brines (squares) (from Richter and others, 1990).

171

may serve as good tracers on a local basis, as these are more concentrated in evaporated ground waters than in most mixing waters between fresh ground water and brine at similar salinities.

3.6.5. State-by-State Summary of Saline Seep Occurrences

Relatively few published reports on saline-seep occurrences in the United States have been found, with the exception of the states of Montana, North Dakota, Oklahoma, South Dakota, and Texas. The information found is summarized in this section.

Montana: Potential for saline-seep development exists in central, northern, and eastern Montana. Saline seep is caused by dryland agriculture and the crop-fallow rotation system necessary for moisture conservation and small-grain production on the scale practiced in the state. In 1969, 51,200 acres had been affected by saline seep (Ferreira and others, 1988). A 1971 survey revealed that approximately 80,000 acres of nonirrigated cropland had been lost to saline seep. In the following four years, an additional 100,000 to 150,000 acres were affected by saline seep (van der Leeden and others, 1975). Affected acreage had increased to 280,000 acres by 1983 and to 380,000 acres by 1984 (Ferreira and others, 1988). Serious conditions have appeared in north-central and northeastern Montana where saline seep is increasing at a rate of over 10 percent a year. Growth and areal extent of saline seep in the Nine Mile watershed illustrate the destruction of land due to saline seep. In 1941 only 0.1 percent of the total land area had been affected by saline seep. The affected area had grown to 0.4 percent in 1951, to 9.1 percent in 1966, and to 19.4 percent in 1971 (Fig. 67). Saline-seep development is most pronounced where the glacial till is less than 30 ft deep. The till is underlain by a thick marine shale that is impermeable to water. Both the till and the shale contain an abundant supply of natural salts that are picked up by excess water and are stored within the perched water table aquifer. Where the water table intersects the land surface, evaporation of the water leads to accumulation of salts. It is estimated, from measurements at Fort Benton, that the water table within the glacial till is rising an average of four to ten inches a year, bringing the water table closer to land surface at topographically low areas (Bahls and Miller, 1975). Ground-water samples are typically high in TDS (some exceeding 50,000 mg/L), sulfate (some exceeding 30,000 mg/L), and nitrate (in the hundreds of mg/L). Community ground-water supplies at Nashua, Wiota, and Frazer have been contaminated by saline seep, and nitrate poisoning of livestock has been reported from the Fort Benton and Denton areas (Bahls and Miller, 1975). Concentrations of greater than 25,000 mg/L TDS (Na, Mg-SO_4 type) have been reported for seep water in the Fort Benton area, one of the first areas to be affected.

Approximately 16 percent (198,000 acres) of the state's irrigable land was considered saline or alkaline in 1960 (Fuhriman and Barton, 1971). Geological conditions that favor development for saline seeps exist over an area of 12,500 mi^2 (van der Leeden and others, 1975). Because of the state's

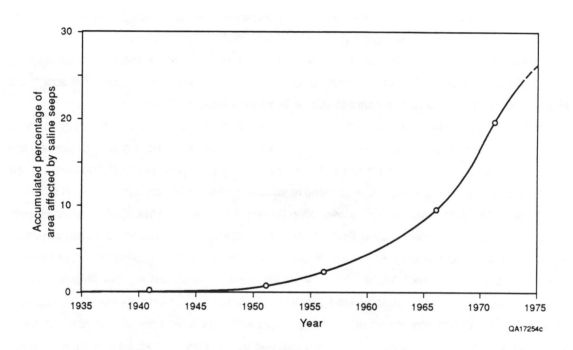

Figure 67. Rate of growth of a saline seep near Fort Benton, Montana, from 1941 through 1971 (from Thompson and Custer, 1976).

position as headwater recharge area of the Missouri River Basin, continued degradation of surface waters by saline seep in Montana can affect many downstream users (Bahls and Miller, 1975).

North Dakota: With the exception of the eastern state-border area, all land areas of North Dakota can be considered potential saline-seep areas (Bahls and Miller, 1975). Miller and others (1980) estimated that saline soils have developed on 20,000 to 40,000 ha of nonirrigated farm land. Of the state's total irrigable land surface, approximately 31 percent (817,000 acres) was considered saline or alkaline in 1960 (Fuhriman and Barton, 1971). In the Red River Valley alone, approximately 400,000 acres of land are affected by excessive salt concentrations (Benz and others, 1961).

Oklahoma: In the 1970's, farmers in Harper County became aware of the effects of saline seep. By 1985, 1,300 ha of about 65,000 ha wheatland in the county were known to be affected. The size of the individual saline seep areas ranged from 5 to 100 ha (Berg and others, 1987).

South Dakota: The major salt-water problem in the state is associated with the occurrence of saline seep (Atkinson and others, 1986), which exists throughout most of the northern, central, and northeastern parts of the state (Bahls and Miller, 1975). More than 70 percent (1,196,000 acres) of the state's irrigable land surface was considered saline or alkaline in 1960 (Fuhriman and Barton, 1971).

Texas: The Texas State Soil and Conservation Board (1984) estimated that approximately 5 million acres of land can be considered saline. Dryland saline seep occurs over large parts of Texas, especially since the 1950's. Approximately 50 percent of the initial saline seep observations fall within the time periods of 1950 to 1959 and 1970 to 1974 (Neffendorf, 1978). The largest number of seeps occurs within the Central Rolling Red Plains, where 1,230 areas of dryland seep cover more than 90,000 acres in 39 counties (U.S. Department of Agriculture, 1983). These saline spots were not evident 75 to 100 years ago. Saline seep areas in that area are generally separated from deeper brine aquifers by impervious soft shale bedrock. This is in contrast to natural saline springs and salt flats in the Rolling Plains, which occur where streambeds and valleys are incised into deep-brine aquifers. Saline spots are also created by oil-field contamination, with a similar visual appearance of being devoid of any vegetation. Levels of salinity, however, are usually higher in oil-field contaminated seeps than in dryland saline seeps (U.S. Department of Agriculture, 1983). The average size of seeps in the Central Rolling Red Plains was 74 acres, with 95 percent identified as increasing in size (U.S. Department of Agriculture, 1983). West of Estelline in Hall County, saline seeps are caused by the removal of natural vegetation and cultivation of the soil (Bluntzer, 1981). These practices and the natural occurrence of an impermeable clay or shale layer has caused the development of perched ground-water conditions, allowing evaporation and salinization at topographically low areas. Richter and others (1990) identified saline seep as a major source of saline ground water in the Runnels County area.

Counties with the largest acreage affected by dryland saline seep in Texas are (approximate acreage in parentheses): Baylor (17,000), Cameron (26,000), Coleman (10,000), Collingsworth (11,000), Foard (13,000), Hardeman (27,000), Haskell (15,000), Hutchinson (19,000), Kleberg (16,000), Knox (14,000),

174

Nueces (10,000), Ward (24,000), Wilbarger (30,000), and Zavala (27,000) (Texas State Soil and Water Conservation Board, 1984).

3.7 Road Salt

3.7.1. Mechanism

Salt has been used as an efficient road deicing agent for a considerable amount of time, with good results regarding its primary purpose of providing safe travel during winter months. Economic benefits of street salting are numerous, such as improved fuel efficiency and reduction of costs associated with accidents, but they don't come without negative side effects. Some of the environmental effects are the contamination of surface runoff, of surface waters, such as lakes and streams, of soils, and of ground water.

The degree of contamination potential of water resources is directly related to the number of years that salt is applied to a given stretch of road and to the amount of salt applied to that stretch each year. More than four million tons of sodium chloride and calcium chloride were used in the United States during the winter of 1966–1967 (Field and others, 1973). This number increased to 10.5 million tons during calender year 1990, representing the approximate average for the past 5 years (Salt Institute, Virginia, personal communication, 1991). Due to regional weather conditions, approximately 95 percent of road-salt usage occurs in eastern and north-central states (Field and others, 1973) (Table 11). Consequently, most water pollution is reported from those states (Table 11), where average application rates often exceed 20 tons of salt per lane-mile (Hutchinson, 1973). The actual amount used depends on weather conditions, such as temperatures and frequency of snowfalls, but also on population numbers and local policies. Increases in population, road networks, and sometimes more aggressive or wasteful application has led to a general increase in salt usage over the years, doubling every five years since 1940 (Fig. 68). It probably can be assumed that the use of salt as a deicing material will continue to fluctuate through the years depending on weather conditions. In the future, however, it will more likely increase than decrease, as more and wider roads are being built. In Massachusetts, an eightfold increase in salt usage occurred between 1954 and 1971 (Miller and others, 1974). This increase in salt application corresponds to an overall increase in chloride concentration in the state's ground water (Fig. 69). State-by-state application rates reported (Fig. 7, Table 11) often mask the local nature of very high usage and very high contamination potentials. For example, Monroe County, New York, alone used between 109,000 and 224,000 tons of salt each winter between 1965–1966 and 1972–1973, which represented approximately 30 percent of New York's total use in 1966–1967 (Diment and others, 1973).

Contamination of water resources can occur from storage piles of salt as well as from application of salt onto roads. Uncovered storage piles may cause some of the highest, local contamination potentials

175

Table 11. Reported use* (in tons) of road salt and abrasives during 1966–1967 (data from Field and others, 1973), during the winters of 1981–1982 and 1982–1983 (data from Salt Institute, undated), and associated known pollution problems listed by state (modified from Geraghty and others, 1973, and miscellaneous sources listed in chapter 3.7.5).

State	NaCl			CaCl$_2$	Contamination
	1966–67	1981–82	1982–83	1966–67	
Arkansas	1,000	2,510	856	-	
California	11,000	13,600	-	-	
Colorado	7,000	22,460	10,896	-	yes
Connecticut	101,000	103,201	51,934	3,000	yes
Delaware	7,000	8,913	7,053	1,000	yes
Idaho	1,000	11,000	11,000	-	
Illinois	249,000	304,184	206,000	10,000	yes
Indiana	237,000	313,365	116,650	6,000	yes
Iowa	54,000	64,000	60,400	2,000	
Kansas	25,000	35,490	31,630	2,000	
Kentucky	60,000	73,275	32,960	1,000	
Maine	99,000	51,676	49,202	1,000	yes
Maryland	132,000	155,758	82,499	1,000	yes
Massachusetts	190,000	262,000	178,500	6,000	yes
Michigan	409,000	397,000	229,000	7,000	yes
Minnesota	398,000	118,587	127,957	14,000	
Missouri	34,000	90,963	75,111	3,000	
Montana	4,000	2,817	3,245	-	
Nebraska	10,000	22,221	24,899	-	
Nevada	4,000	8,500	9,831	-	
New Hampshire	118,000	138,692	93,813	-	yes
New Jersey	51,000	138,692	35,700	6,000	yes
New Mexico	7,000	16,000	23,000	-	
New York	472,000	443,000	300,000	5,000	yes
North Carolina	17,000	45,264	36,573	2,000	yes
North Dakota	2,000	8,222	8,719	1,000	yes
Ohio	511,000	401,285	184,341	12,000	yes
Oklahoma	7,000	9,300	18,770	-	
Oregon	1,000	-	456	-	
Pennsylvania	592,000	500,010	231,000	45,000	yes
Rhode Island	47,000	56,280	29,297	1,000	yes
South Dakota	2,000	4,345	3,697	1,000	
Texas	3,000	-	-	-	
Utah	28,000	79,540	79,720	-	
Vermont	89,000	71,904	65,647	1,000	yes
Virginia	77,000	178,500	95,000	22,000	yes
Washington	2,000	10,000	7,500	-	
West Virginia	55,000	90,636	52,709	9,000	yes
Wisconsin	225,000	236,790	229,803	3,000	yes
Wyoming	1,000	5,000	6,340	-	

* States not included are due to unavailability of data.

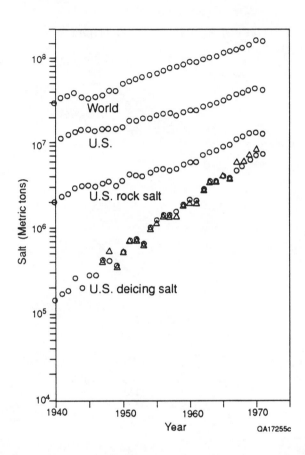

Figure 68. Salt production and salt usage for deicing. The use of salt for road deicing increased steadily between 1940 and 1970 at a rate that exceeds that of the increase in production of salt from brine, evaporation, and rock salt (from Diment and others, 1973).

177

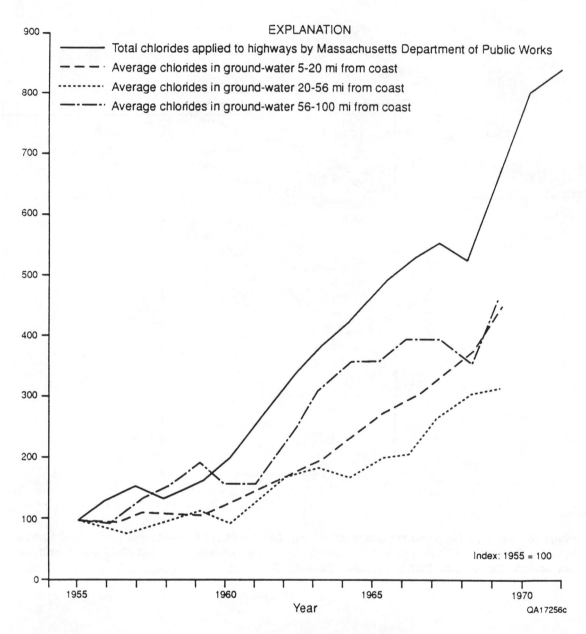

Figure 69. Correlation between increase in salt usage applied to highways and chloride concentrations in ground water, Massachusetts, 1955–1971 (from National Resources and Agriculture Committee, 1973; Miller and others, 1974).

because of the concentrated nature of brine originating at those piles after precipitation events. Such a case was described by Wilmoth (1972) for Monroe County, West Virginia, where brine seeped into a fractured carbonate aquifer. There it mixed with fresh water, and subsequently appeared in a water well located approximately 0.3 mile away from the storage pile. The number of storage piles goes into the thousands, with each holding several hundreds to thousands of tons of salt every fall (Miller and others, 1974). These storage sites are also often used as wash stations of salt-spreading trucks, with drainage from salt piles and the wash areas being fed into dry wells (Miller and others, 1974). Road spreading of salt may lead to contamination during the spreading phase or during the runoff phase. If the salt is applied as a powder, salt particles may become airborne and transported considerable distances downwind. Similarly, if the salt is applied as a brine solution, a fine spray or mist can be transported by wind (Jones and Hutchon, 1983). Melting of ice and snow on the road creates brine solutions which enter drainage ditches or roadside fields. Extensive drainage-ditch networks in urban areas may catch most of the saline surface runoff. This is the case in the city of Buffalo, where an estimated 90 percent of the applied road salt ends up in the city's sewer system (Rumer and others, 1973). In rural areas, removal of salt by surface runoff may be less efficient (for example, 50 percent, Bubeck and others, 1971), with the remaining salt being flushed into soil and then into ground water. While drainage ditches can divert the salt problem away from highways, such as in the case of Buffalo, which discharges its water into the Niagara River (Rumer and others, 1973), runoff into unlined ditches or onto fields concentrates the problem along highways. The pollution problem generally decreases with increasing distance from the roads, but as the salt content in soil and ground water increases over the years the extent of contamination may increase (Hutchinson, 1973). Miller and others (1974) reported pollution several thousand ft away from sources of salt, with a penetration to depths of almost 400 ft in a few wells. This could cause problems for wells typically located near roads for easy accessibility. Wilmoth (1972) estimated that more than half of all the water wells in West Virginia are located within 100 to 500 ft of salt-treated roads which would place them relatively close to the area of potential contamination at present or sometime in the future. Similar conditions probably exist throughout most of those areas where road salt is used.

3.7.2. Road-Salt Chemistry

Very few complete chemical analyses of road-salt brines or road-salt-affected water have been published. Where endmember chemistry of salt solutions are of interest for statistical evaluation or for mixing calculations, some researchers have used analyses of halite solutions prepared in the laboratory. This was done, for example, by Novak and Eckstein (1988), who reported a Na-Cl water with very low concentrations of other major chemical constituents. Molar concentrations of sodium and chloride in those analyses are equal, reflecting dissolution of nearly pure halite. Street salt is not pure halite in many instances, however, but instead consists of a mixture of sodium chloride and calcium chloride, with sometimes relatively high concentrations of potassium and boron and low concentrations of sulfate.

3.7.3. Examples of Geochemical Studies of Road Salting

Water wells along highways in Maine showed increases from background levels of 3–4 ppm in sodium and chloride to 70–76 ppm average sodium and 150–171 ppm average chloride caused by street salting. The increase in sodium lags behind the increase in chloride because sodium ions are retained in the soil whereas chloride ions enter the ground-water system unrestricted (Hutchinson, 1973). Chloride concentrations were also used by Diment and others (1973) to assess surface-and ground-water contamination in Monroe County, New York. Although two other sources of high chlorinity, such as sewage and naturally saline formation water, were present in Monroe County, chloride concentrations were used successfully after water-budget considerations had identified these sources as minor contributors to the salinity problem.

In two case studies reported by Aulenbach (1980), brine seepage from salt storage piles in New York was suspected to contaminate water wells. Due to the highly permeable nature of the affected aquifers—fractured limestone in one case, gravel in the other case—rhodamine dye was used successfully to positively identify the salt-storage piles as the point sources of pollution. The tracer was detected in the affected wells after 2 weeks and 3 months, respectively.

Novak and Eckstein (1988) used discriminant analysis and constituent ratios between major cations and chloride to distinguish water possibly contaminated by road salt from water possibly contaminated by oil-field brine. The rationale for using only the constituents Ca, Mg, Na, K, and Cl was that this allowed inclusion of older analyses, which did not report SO_4, HCO_3, or any minor constituents and because emphasis was put on constituents that are determined routinely and inexpensively. Using contaminated ground water from Perry Township, Ohio, as test samples, oil-field brines and salt solutions as the two potential endmembers of salinization, fresh ground water as the third endmember, and the ion ratios Ca/Mg, Na/Ca, Na/Mg, Na/Cl, K/Cl, K/Na, Mg/K, Ca/K, Cl/Mg, Cl/Ca, and (Ca+Mg)/(Na+K), Novak and Eckstein (1988) were able to eliminate street salt as a possible source. Discriminant analysis grouped the contaminated waters as more similar to oil-field brines or to fresh ground water than to salt solution. Snow and others (1990) followed the approach of Novak and Eckstein (1988) in using modified Stiff diagrams (with constituent ratios instead of concentrations as end points) to distinguish sea-water intrusion from road-salt contamination in coastal wells of Maine. In addition to graphical display in Stiff diagrams, these sources of salinity were best distinguished using bivariate constituent plots of Br versus Cl and of SO_4 versus Cl. Road-salt contamination was characterized by Br/Cl ratios and SO_4/Cl ratios being substantially lower than in samples affected by sea-water intrusion (Snow and others, 1990).

Knuth and others, 1990, differentiated road-salt contamination in Ohio (location not specified) from deep-formational brine disposed of in surface pits during drilling of a gas well and naturally occurring saline water in the Meadville Formation by analyzing Br and Cl concentrations in fresh, uncontaminated water, in gas brine, and in contaminated water wells. Construction of mixing curves on a Br/Cl versus Cl plot identified two wells (Fig. 70), which had experienced increases in chloride concentrations of only several mg/L to 70 mg/L and which displayed low Br/Cl ratios (14×10^{-4} to 60×10^{-4}), as affected by road salt (end-member Br/Cl ratio of approximtely 1×10^{-4}). Br/Cl ratios of approximately 100×10^{-4} helped to identify two wells as affected by gas brine (end-member Br/Cl ratio of approximately 115×10^{-4}), whereas wells affected by upward flow of naturally saline ground water displayed intermediate Br/Cl ratios (50×10^{-4} to 100×10^{-4}).

3.7.4. Significant Chemical Parameters

By far the most widely used parameter in identification of street-salt contamination is the chloride ion. Chloride is a good tracer because it is the most conservative ion dissolved in ground water, it is the most abundant ion in street-salt solutions, and it is analyzed on a routine basis. Background chloride concentrations are known for a vast number of water wells all over the country. Because contamination from street salt is a seasonal phenomenon with high chloride concentrations in spring runoff and decreasing (dilution) concentration throughout the remainder of the year, deviation of chloride concentrations from background levels are a good measure of the degree of salt contamination in most instances. Accumulation of salt may occur in the soil and in ground water, which means that background levels may increase over the years. When salt-brine runoff infiltrates the vadose zone and the saturated zone, sodium is often absorbed to soil and aquifer material. Therefore, the Na/Cl ratio may be smaller in salt-affected ground water than in salt-affected surface water.

Because of its conservative nature once dissolved in ground water, bromide can be a good tracer of salinity. Expressed as Br/Cl weight ratios, it can be used to differentiate salinity derived from road salt (halite) as opposed to oil- and gas-field brines, deep-formation waters, and sea water, as halite solution produces some of the lowest Br/Cl ratios measured in naturally saline waters.

On a local basis, high concentrations of calcium and chloride may be indicative of road-salt contamination where large amounts of $CaCl_2$ are added to the salt mixture.

Under certain circumstances, dye tracer (rhodamine) may be useful for identifying point sources of alleged street-salt contaminations.

The literature review identified very few complete chemical analyses of ground water affected by road salt. This may indicate that the parameters discussed are sufficient for identification of road-salt contamination in most instances.

181

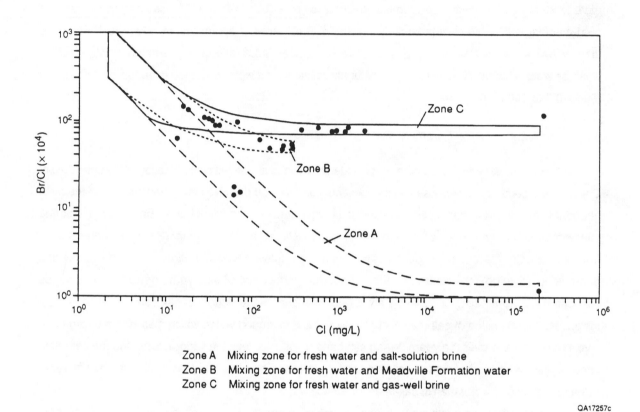

Zone A   Mixing zone for fresh water and salt-solution brine
Zone B   Mixing zone for fresh water and Meadville Formation water
Zone C   Mixing zone for fresh water and gas-well brine

QA17257c

Figure 70. Bivariate plot of Br/Cl ratios versus Cl for selected water samples from northeastern Ohio. Mixing curves delineate salinization by road salt (zone A), by deep saline ground water (zone B), and by gas-field brine (zone C) (from Knuth and others, 1990).

3.7.5. State-by-State Summary of Road-Salt Issues

The amounts of road salt used by and listed for individual states below, in Table 11, and in figure 7 represent the tonnage of salt applied mostly by state highway departments, as reported during salt surveys. These numbers should be considered conservative because they do not include salts applied by towns and cities with their own deicing programs and because not all agencies respond to such surveys.

Connecticut: Approximately 82,000 tons of road salt were used in the state during the winter of 1965–1966, which translated to 9 tons per single-lane mile. In the following winter (1966–1967), road-salt application increased to 104,000 tons (Field and others, 1973), the same rate which was applied 10 years later during the winter of 1976–1977 (Bingham and Rolston, 1978). The large variability of salt usage is reflected in the amounts of salt used during the winters of 1981–1982 and 1982–1983, reported at 103,201 and 51,934 tons, respectively (Salt Institute, undated). High chloride and sodium-ferrocyanide concentrations in some water supplies have been traced to salt-storage areas in the state (Scheidt, 1967; in Field and others, 1973).

Illinois: The reported use of sodium and calcium chloride as road-deicing agents during the winter of 1966–1967 amounted to 259,000 tons (Field and others, 1973). This is a 58 percent increase from the previous year (164,000 tons), during which already more than 20 tons of salt were applied per lane mile (Geraghty and others, 1973). During the winters of 1981–1982 and 1982–1983, salt usage varied from 304,000 to 206,000 tons, respectively (Salt Institute, undated).

Contamination by road salt is indicated by an abrupt increase in salt content during the major spring thaws with smaller subsequent increases. Minimum salt increases are generally observed in late fall and early winter. Ground-water contamination is less abrupt but shows a consistent increase in yearly minimum chloride levels when compared to levels recorded prior to heavy street-salt usage (Walker, 1970). In Peoria, contamination of several water wells was caused by leakage of brine from a salt-storage pile into an old storm sewer and subsequent leakage from the sewer into shallow ground water (Walker, 1970).

Maine: Approximately 100,000 tons of sodium and calcium chloride were applied in Maine during the winter of 1966–1967 (Field and others, 1973). Usage was only about half that amount during 1981–1982 and 1982–1983 (Salt Institute, undated). The average annual application rate of road salt for the past several years (1973 date) has been 25 tons per two-lane mile (Hutchinson, 1973). This application has led to increased salt concentrations in ground water in many areas of the state. Sodium and chloride levels in soils next to highways exhibit a direct relationship with the number of years salt has been applied. In areas where salt had been applied for 20 years, the sodium levels rose over a distance of 60 ft from the shoulder and to a depth of 18 inches. However, surface waters have not been markedly affected owing to the high degree of dilution of saline solutions (Hutchinson, 1973).

Seasonal variation of chloride concentrations in some water wells in Maine suggest contamination by road salt. While chloride content in various aquifers in the state is normally less than 20 mg/L, the three-year (1967–1969) average April chloride content of water from approximately 100 sampled wells was 171 mg/L; the highest concentration of sodium was 846 mg/L and that of chloride was 3,150 mg/L. Concentrations were less in August than in April, which is the month of greatest snow melt and runoff from highways (Miller and others, 1974).

Maryland: Approximately seven tons of road salt were applied to each single-lane mile during the winter of 1965–1966, for a total of 45,000 tons (Miller and others, 1974). By the next winter (1966–1967), this amount had tripled to approximately 133,000 tons of salt (Field and others, 1973), similar to the amount used during the winter of 1981–1982 (Salt Institute, undated).

Massachusetts: During the winter of 1965–1966, approximately 126,000 tons of sodium and calcium chloride were applied as road-deicing agents, which is equivalent to nearly 21 tons per single-lane mile (Miller and others, 1974). Salt use increased during the winter of 1966–1967 to 196,000 short tons, during the winter of 1981–1982 to 262,000 tons, and during the winter of 1982–1983 to 178,500 tons (Salt Institute, undated). Many occurrences of water-supply contamination due to road salt have been recorded in the state. Among them are the town of Burlington, the town of Becket, Mystic Lakes, and wells in the Weymouth, Braintree, Randolph, Holbrook, Auburn, Tyngsboro, Charlton, and Springfield areas (Field and others, 1973).

Michigan: During the winter of 1966–1967, approximately 416,000 short tons of sodium and calcium chloride were applied throughout Michigan for deicing of roads. During the wnter of 1981–1982, salt usage was only slightly lower (397,000 tons), but during the winter of 1982–1983, salt usage dropped to 229,000 tons (Salt Institute, undated). Contamination from salt-storage piles has occurred in Manistee County, where a well located 300 ft away from a storage pile contained chloride concentrations of up to 4,400 mg/L (Schraufnagel, 1967). An unprotected salt-storage pile also may have contaminated portions of the Black River limestone near the village of Rock (Moore and Welch, 1977). The Michigan Department of Natural Resources (1982) identified 33 known and 86 suspected incidents of salt-storage or road-salt contamination cases for the 1979 calendar year. Known contamination, mostly from salt-storage areas, occurred in 19 counties. In an additional 29 counties contamination was suspected. Of the total 203 road-salt storage facilities, more than 50 percent had known or suspected contamination problems during that year.

Minnesota: Approximately 412,000 tons of sodium and calcium chloride were used during the winter of 1966–1967 (Field and others, 1973), an amount significantly above that used during the winters of 1981–1982 (118,587 tons) and 1982–1983 (127,957 tons) (Salt Institute, undated). The average application rate is 15 tons per highway mile (Schraufnagel, 1967, Hutchinson, 1973).

New Hampshire: During the winter of 1965–1967, approximately 12 tons of road salt were applied to every single-lane mile in New Hampshire. This totals more than 83,000 tons of sodium and calcium chloride salt (Miller and others, 1974). The following winter, approximately 118,000 tons of road salt were used. In the early 1970's, application rates had increased to an average of 150,000 tons per year (Morrissey and Regan, 1988). The number of wells being abandoned each year as a result of street-salt contamination increased from four wells in 1953 to 37 wells in 1964 (Moore and Welch, 1977). By 1965, more than 200 road-side wells had to be abandoned due to contamination by road salt; chloride concentration exceeded 3,500 mg/L in some of those wells (Field and others, 1973). As of January 1987, approximately 79 percent of all contaminated wells in the state were contaminated by road salt (Morrissey and Regan, 1988). This high number of abandoned and contaminated wells may be caused by the accumulation of salt over the past 50 years, during which more than 4 million tons of salt were used within the relatively small area of this state.

New York: During the winter of 1965–1966, approximately 250,000 short tons of sodium chloride and calcium chloride were used in New York. This is equivalent to 7.5 tons per single-lane mile (Miller and others, 1974). By the winter of 1966–1967, salt usage had increased to approximately 477,000 short tons (Field and others, 1973). Between 1965 and 1973, the use of deicing salt in Monroe County (Lake Ontario) alone varied between 109,000 and 224,000 metric tons per winter. This has caused (a) a fourfold rise of chloride concentration in Irondequoit Creek and Irondequoit Bay waters, (b) maximum chloride concentrations in creeks ranging from 260 to 46,000 mg/L during winter season, (c) a decrease in mixing of bay waters due to density stratification, and (d) possible increases in chloride concentrations in water wells. Approximately 50 percent of the salt that is used is removed by surface runoff. The other 50 percent are stored in soils and ground water (Diment and others, 1973).

Aulenbach (1980) reports of the successful use of Rhodamine WT dye for tracing contamination of water wells by nearby salt-storage areas in New York. In the two cases, the dye, which had been distributed around the storage areas, reached the domestic wells after two weeks in one and after three months in the other. Chloride concentrations in those wells had increased to greater than 1,700 mg/L as a result of salt-water seepage from the uncovered salt-storage piles.

Approximately 23,000 tons of deicing salt are discharged annually through sewer systems into Lake Erie by the City of Buffalo (Rumer and others, 1973).

North Carolina: During the winter of 1966–1967, approximately 19,000 tons of sodium chloride and calcium chloride were used throughout the state (Field and others, 1973). About the same amount, 18,000 tons, was used during the winter of 1973–1974 (Miller and others, 1977), but usage increased substantially during the winters of 1981–1982 (45,264 tons) and 1982–1983 (36,573 tons) (Salt Institute, undated). In Haywood County a stockpile of salt caused contamination of a water well which was indicated by a maximum chloride content of 1,320 mg/L in the well water (Miller and others, 1977).

Ohio: During the winter of 1966–1967, approximately 523,000 short tons of sodium and calcium chloride were used as road salt throughout the state (Field and others, 1973). This one application of salt has caused local salinization of ground and surface water. Chloride concentrations increase downstream from storm-drain outlets in the Olentangy River near Columbus (Kuhlman, 1968). Knuth and others, 1990, differentiated road-salt contamination in Ohio (location not specified) from deep-formational brine disposed of in surface pits during drilling of a gas well and naturally occurring saline water in the Meadville Formation by analyzing Br and Cl concentrations in fresh, uncontaminated water, in gas brine, and in contaminated water wells. Construction of mixing curves on a Br/Cl versus Cl plot identified two wells, which had experienced increases in chloride concentrations of only several mg/L to 70 mg/L and which displayed low Br/Cl ratios (14×10^{-4} to 60×10^{-4}), as affected by road salt (endmember Br/Cl ratio of approximately 1×10^{-4}) (Fig. 70). Br/Cl ratios of approximately 100×10^{-4} helped to identify two wells as affected by gas brine (endmember Br/Cl ratio of approximately 115×10^{-4}), whereas wells affected by upward flow of naturally saline ground water displayed intermediate Br/Cl ratios (50×10^{-4} to 100×10^{-4}).

Pennsylvania: During the winter of 1966–1967, approximately 637,000 short tons of sodium and calcium chloride were used as road salt throughout the state (Field and others, 1973). This is a 136 percent increase compared to the previous year (270,000 tons; average of more than 20 tons per lane-mile; Geraghty and others, 1973), translating to an average application rate greater than 50 tons per lane-mile. Application rates were much lower, however, during the winters of 1981–1982 (500,000 tons) and 1982–1983 (231,000 tons) (Salt Institute, undated).

South Carolina: Only about 50 tons of road salt were used throughout the state during the winter of 1973–1974 (Miller and others, 1977). Usage increased to 3,450 tons during the winter of 1981–1982 and 3,697 tons during the winter of 1982–1983 (Salt Institute, undated).

Vermont: Deicing salts are major sources of elevated sodium and chloride concentrations in some areas of the state; more than 30 percent of the wells that produce water from contaminated aquifers are contaminated by salt from road application or storage (Cotton and Butterfield, 1988). During the winter of 1965–1966, approximately 84,000 short tons of road salt (sodium chloride and calcium chloride) were applied throughout the state, averaging approximately 18 tons per single-lane mile (Miller and others, 1974). Only slightly more than that, approximately 90,000 short tons (Field and others, 1973) were used during the winter of 1966–1967. Most of this salt may end up in surface streams, as documented by Kunkle (1971) for a watershed that received 63 to 100 tons of salt per year and had a calculated 83 ton/year excess of NaCl when compared to background levels.

Virginia: During the winter of 1966–1967, approximately 77,000 short tons of sodium chloride and 22,000 short tons of calcium chloride were used in Virginia (Field and others, 1973). Application rates were less during the winter of 1973–1974, amounting to approximately 63,000 short tons (Miller and others, 1977), but more than twice that much (178,500 tons) during the winter of 1981–1982 (Salt Institute, undated). Contamination of water wells by salt-storage areas has been reported by Miller and others (1977) for Goochland and Dinwiddie Counties.

West Virginia: During the winter of 1966–1967, approximately 64,000 short tons of sodium chloride and calcium chloride were applied as road salt throughout the state (Field and others, 1973).

186

Incidents of deteriorating water quality due to road-deicing salts are increasing every year. Chloride concentrations of more than 10,000 mg/L have been observed in runoff from heavily salted areas of Grant and Raleigh Counties. It is estimated that more than half of all the water wells in the state are located within 100 to 500 ft of roads that receive some salt treatment. Several hundred tons of road salt stored unprotected have caused well contamination near Union, Monroe County. In the absence of any other salt source, chloride concentrations in road-salt affected wells increased from about 25 mg/L to as high as 7,200 mg/L (Wilmoth, 1972).

Wisconsin: During the winter of 1966–1967, approximately 228,000 tons of road salt were applied throughout the state (Field and others, 1973). Similar amounts were used during the winters of 1981–1982 (236,790 tons) and of 1982–1983 (229,803 tons) (Salt Institute, undated). The average application rate of road salt in the state is 15 tons per mile (Schraufnagel, 1967; Hutchinson, 1973). Chloride concentrations in surface runoff from highways as high as 10,250 mg/L have been measured in parts of the state. This caused seasonal changes in surface-water concentrations from background levels of 0.5 to 2 mg/L dissolved chloride to maximum levels of 45.5 mg/L (Moore and Welch, 1977). Chloride concentrations in lakes that receive highway-salt runoff may not be uniform but stratified, as reported for Beaver Dam Lake by Schraufnagel (1967). Chloride concentrations in this lake increased from 8 mg/L at the top to 33 mg/L at the bottom at 15 ft.

4. GEOCHEMICAL PARAMETERS

This chapter will provide a brief introduction to characteristics of individual parameters that have been used in salinization studies in the past, and that may have been selected for a particular problem study during the previous chapters. Also included are references to field and laboratory methods and approximate analytical costs.

It is absolutely crucial in a salinization study to know which methods and parameters are the best to use for a particular problem. Time and money may play an important role, but technology does also, as better methods are available now than were available in the past, and still better methods will be available in the future.

Through the years, a variety of chemical constituents and constituent ratios have been used as possible tracers of salinity sources (Table 12). Parameters most often used include the major cations Ca, Mg, Na, the major anions HCO_3, SO_4, Cl, some minor elements (K, Br, I, Li), and some isotopes (^{18}O, ^2H, ^3H, ^{14}C). Some of these constituents are more useful than others, as discussed in the following sections that deal with individual chemical and isotopic constituents listed in Table 12. Constituents are listed in alphabetical order.

4.1. Discussion of Individual Parameters

Aliphatic Acids: High concentrations of short-chain aliphatic acids (acetate, propionate, butyrate, valerate) in some oil-field/deep-basin brines give way to very low or zero concentrations in low-saline mixing waters. This is due to dilution and progressive biodegradation of the organic acids (Hanor and Workman, 1986, Kreitler and others, 1990). Therefore, their usefulness in salinization studies is limited to differentiation of little-diluted endmember brines.

Alkalinity: Alkalinity represents the capacity of a solution to neutralize acid. It is determined in the field by titration of a sample aliquot (50 to 100 ml are sufficient in most instances) to an endpoint pH of approximately 3.0 using a strong acid ($6N$ H_2SO_4) (Brown and others, 1970; Wood, 1976). Alkalinity measurements in the laboratory are typically slightly lower than actual values (Roberson and others, 1963). At the ph range of most natural waters, alkalinity is represented mainly by the dissolved carbon dioxide species HCO_3^- and CO_3^{2-}; commonly, alkalinity is reported as bicarbonate (HCO_3^-) or calcium carbonate ($CaCO_3$). To convert alkalinity expressed as calcium carbonate to alkalinity expressed as bicarbonate, the amount (in mg/L) of calcium carbonate is multiplied by the factor of 1.22. Alkalinity is generated by the action of dissolved atmospheric carbon dioxide and soil carbon dioxide in the water on carbonate rocks, such as limestone. Respiration by plants and the oxidation of organic matter in the soil and in the unsaturated zone increase CO_2 content over atmospheric concentrations (0.03 percent). Additional sources of carbon dioxide result from biologically mediated sulfate reduction, metamorphism of

Table 12. Geochemical parameters used for identification of salinity sources.

Salinization Sources	Chemical Parameter	Page
Natural saline water versus others	Cl	28
	Br, I, S-34, ^{18}O, D, Br/Cl, Na/Cl, I/Cl, I	28
	Mg/Cl, K/Cl, Ca/Cl, (Ca+Mg)/SO$_4$, Sr	28
Halite-solution brine versus others	K/Na, Br/TDS	62
	(Ca+Mg)/(Na+K), Na/Cl,	62
	Ca/Cl, Mg/Cl, SO$_4$/Cl,	66
	Br/Cl	66
	K/Cl, (Ca+Mg)/SO$_4$, I/Cl	68
	^{18}O/D, I/Cl, SO$_4$/(Na+K), SO$_4$/TDS, SO$_4$/Cl	68
Sea-water intrusion versus others	Cl	88
	Major ions (Piper)	92
	^{14}C, ^{3}H	96
	I/Cl, B, Ba, I	98
	^{18}O, ^{2}H, ^{13}C	98
	Ca/Mg, Cl/SO$_4$, B/Cl, Ba/Cl	100
	Br/Cl	100
Oil-field brines versus others	Cl, Major ions	123
	Na/Cl	128
	Ca/Cl, Mg/Cl, SO$_4$/Cl, Br/Cl	128
	I/Cl, Major ion ratios, Cl, Br, (Na+Cl)/TDS,	131
	Li/Br, Na/Br, Na/Cl, Br/Cl	132
Agricultural effluents versus others	Cl, NO$_3$, Cl/NO$_3$	149
	K, TDS	149
Saline seep versus others	SO$_4$	163
	Ca/Cl, Mg/Cl, SO$_4$/Cl	168
	NO$_3$	170
Road salt versus others	Cl	180
	Major ion ratios, Br/Cl	180
	Dye	181

carbonate rocks, and outgassing from rocks in the Earth's mantle (Hem, 1985). In some instances, especially in oil-and gas-field associated waters, short-chain aliphatic acids may contribute to total alkalinity of the water sample and sometimes have been confused with inorganic alkalinity. Decarboxylation of short-chain aliphatic acids (for example, acetic acid) in oil and gas fields may also contribute large amounts of carbon dioxide (Carothers and Kharaka, 1978).

In most natural waters alkalinity ranges from a few tens of mg/L to <1,000 mg/L. Low concentrations occur in soils and rocks low in calcium carbonate, whereas high concentrations are reported in water flowing through carbonate-rich soils and aquifers; very high concentrations can be found in sandstone aquifers as Na-HCO$_3$ waters. Concentrations >1,000 mg/L have been reported, especially in waters associated with oil and gas reservoirs. In the latter case, other constituents, such as short-chain aliphatic acid anions (for example, acetate, valerate) may contribute to or represent the bulk component of alkalinity. Extremely high alkalinity concentrations (as HCO$_3$) of close to 20,000 mg/L have been reported by Kaiser and others (1991) in deep ground water associated with coal deposits in the San Juan Basin, New Mexico. According to these authors, the high inorganic-alkalinity concentrations are bacterially derived through CO$_2$-releasing processes, such as sulfate reduction and methanogenesis. The low concentrations of calcium and magnesium in the same waters prevent removal of dissolved HCO$_3$ from solution through precipitation of carbonate minerals.

Concentrations of bicarbonate are usually much higher (several hundred mg/L) in fresh waters than in oil-field brines (several tens of mg/L), making it a potential tracer of oil-field brine pollution. However, care must be exercised not to compare bicarbonate in fresh water with total alkalinity in oil-field brine because titration of total alkalinity will include any short-chain aliphatic acids present in the brine. Bicarbonate concentrations depend strongly on pH and CO$_2$ partial pressure and therefore, are easily changed by a change in chemical environment along the flow path of ground water. This makes bicarbonate a marginal tracer of salt-water sources although it has been used as constituent ratio of Cl/HCO$_3$ in combination with other ratios of chemical constituents to identify salinization trends (for example, Collins, 1969; Burnitt and others, 1963).

Argon: Argon is produced by radioactive decay of potassium-40. Its concentration dissolved in ground water is dependent on the temperature of the recharging water and thus could allow differentiation between salt waters originating in different geographic areas and differentiation between continental saline water and sea-water intrusion (Custodio, 1987). No salinization studies using argon have been conducted in the United States, limiting the usefulness of argon as a tracer of salinization sources because of the lack of data and documentation of its usefulness.

Bromine: Bromine, present in water as the bromide ion Br⁻, is generally a very good tracer of salinization sources in combination with chloride. Both constituents are relatively conservative, that is, once in solution they are not easily removed by processes such as ion exchange (because of their large size) or precipitation (because of high solubility). High concentrations occur in sea water (65 mg/L), many geothermal waters (several tens of mg/L), and in deep-basin brines and oil-field brines (several tens to greater than 2,000 mg/L). Most saline waters encountered in salinization studies are undersaturated with respect to halite, that is, the Br/Cl ratio in a mixing water is not affected by mineral precipitation but reflects contributions by the principal salinization source(s). The ratio of Br/Cl (or Cl/Br) is especially well suited to distinguish halite-dissolution brine from oil-field/deep-basin brines because only small amounts of bromide are incorporated into the crystal structure of halite during evaporation (distribution coefficient <1.0). When this halite crystal is dissolved by fresh water, the resulting Br/Cl ratio in solution will be small. The Br/Cl ratio is typically one or more order of magnitude smaller in halite-dissolution brines (Br/Cl <5 $\times$ 10^{-4}) than in oil-field/deep-basin brines. In brines that have gone through a halite-reprecipitation stage, bromide concentrations are high because proportionally more chloride than bromide was incorporated into the halite deposits, rendering the final solution higher in Br/Cl than the initial solution. Continued dissolution and recrystallization of halite further lowers Br/Cl ratios in the halite and increases Br/Cl ratios in the solution. Therefore, a brine that originated by dissolving halite and which later went through a halite-reprecipitation stage may show relatively high Br/Cl ratios (Land, 1987), such as basinal brine, which would then have a Br/Cl ratio very different from a halite-dissolution brine that originated just by dissolving halite. Salinization derived from mixing of fresh water with sea water or sea spray is also generally recognized by higher Br/Cl ratios (34 $\times$ 10^{-4}) than are halite-solution brines (typically <10 $\times$ 10^{-4}).

As is the case for all tracers, the beneficial nature of Br/Cl ratios as tracers of salinization sources decreases when the ratios are similar to both endmembers under consideration. That is, oil-field brines from different reservoirs may have Br/Cl ratios too similar to allow distinction between them. Also, differentiation of salt-water sources using Br/Cl ratios works best at high concentrations of TDS. Many investigators have made use of the Br/Cl (Cl/Br) ratio in their studies, including Collins, 1969, Patterson and Kinsman, 1975, Whittemore and Pollock, 1979, Kreitler and others, 1984, 1990, Morton, 1986, Richter and Kreitler, 1986a,b, and Richter and others, 1990. The conservative nature of bromide and chloride ions have also been used to distinguish the mixing of fresh ground water with street runoff as opposed to the mixing of fresh ground water with sewage effluent, as street runoff is typically relatively high in bromide content (Behl and others, 1987).

The chemical character of bromide and the relatively low cost of bromide determination (Table 13) in the laboratory make this an excellent constituent to be used in salinization studies. Whittemore (1988) and Banner and others (1989) warned, however, that many Br/Cl ratios based on data from the early literature may be in error because of past difficulties with bromide determination in the laboratory.

Calcium: Calcium is the most abundant alkaline-earth metal and is a major component of minerals in most aquifer types. The concentration of calcium in ground water is governed by the abundance of calcium-bearing minerals in the aquifer material, equilibrium conditions between the solid, solution, and gas phases, and the presence of minerals with high cation-exchange capacities (Hem, 1985). Calcium concentrations in ground water are high in aquifers consisting of limestone, dolomite, gypsum, or

Table 13. Approximate costs of chemical and isotopical analyses of constituents covered in this report, as reported by various laboratories (no specific laboratories are recommended).

Constituent	($) Cost per sample
Na,K,Mg,Ca,Li,SiO$_2$,B,Ba (ICP-OES)	32.00
Cl (Ion Chromatography)	10.00
SO$_4$ (do.)	19.00
Br (do.)	10.00
NO$_3$ (do.)	20.00
I (Spectrophotometry)	10.00
Br (do.)	10.00
pH	10.00
Alkalinity	10.00
TDS	5.00
TOC	20.00
Oxygen-18 and deuterium	80.00
Tritium	280.00
^{14}C	225.00–440.00
^{34}S	85.00

gypsiferous shale. In those environments, calcium concentrations are typically much higher than chloride concentrations. Addition of salt water, normally much higher in chloride concentrations than in calcium concentrations, from one source can be detected using the change in ion ratios such as Ca/Cl. However, if more than one source of high chlorinity is suspected or if a high-Ca brine is a suspected endmember of mixing, the ratio of Ca/Cl may be of little use to differentiate between these sources (Leonard, 1964). Ion exchange on clay minerals or albitization of plagioclase or K-feldspar can enrich waters in calcium at the cost of sodium (Banner and others, 1989). Because calcium concentrations are highly dependent on pH, partial pressure of CO_2, and the availability of carbonate minerals, the amount of calcium in ground water is more often related directly to chemical processes such as precipitation, dissolution, and ion exchange than to simple mixing of fresh water with saline water.

Calcium analysis is performed in the laboratory at a modest cost together with other major and minor cations (Table 13). Also, sampling is uncomplicated (see Major Ions, this chapter), which justifies determination of calcium as a standard technique in water-quality studies.

Chloride: More than three-quarters of the total amount of chloride in the Earth's outer crust, atmosphere, and hydrosphere is contained in solution as Cl^- ion in the oceans (Hem, 1985). This reflects the chemically conservative nature of chloride, that is, once in solution, it is not easily removed by processes other than precipitation at very late evaporation stages. Concentration in rain water may vary from 1–20 mg/L at the coast to less than 1 mg/L further inland. Ocean spray and wind transport may also cause high chloride concentrations in coastal surface and ground water. Most fresh-water sources contain chloride in the mg/L to the tens of milligrams per liter range, the upper limit for drinking water being 250 mg/L (U.S. Environmental Protection Agency, 1977). Concentrations commonly increase with depth and with distance from recharge areas, approaching the concentration of sea water (approximately 19,000 mg/L) or greater than that in many sedimentary basins. Deep formation waters and especially waters associated with oil and gas often contain chloride concentrations in excess of 100,000 mg/L. Major natural sources of chloride in ground water are dissolution of late-stage evaporites (NaCl), flushing of saline waters retained in predominantly fine-grained sediments since deposition, and sea-water intrusion. Anthropogenic sources of chloride in ground water include highway deicing salts, industrial, domestic, and agricultural wastes, oil-field brines, and pumping-induced salt-water intrusion, including sea-water intrusion. Evaporation of water, naturally and human-induced, increases chloride concentrations in surface and ground water. As a direct result of its conservative nature and involvement in nearly all processes of salinization of ground water, chloride is the most-often used parameter to identify deterioration of water quality. Most studies dealing with salinization due to street-deicing salt or with sea-water intrusion make use of chloride mapping to define the extent of salt-water movement. In general, because of its relatively conservative nature once in solution (see also discussion of Bromine, this chapter), chloride concentrations are excellent tracers of single salt-water sources. A wide combination of constituent ratios of X/Cl (X = Ca, Mg, Na, K, SO_4, HCO_3, Br, I, Li) is used to distinguish between two or more salt-water sources, such as sea water, oil-field brine, halite-dissolution brine, connate water, or sewage. Chloride is determined in the laboratory by titration, using a sample aliquot from the anion sample bottle, or by ion chromatography (Table 13).

Iodine: Iodine is being used successfully as geochemical indicator for gas and oil reservoirs due to its close association with volatile fatty acids (Carothers and Kharaka, 1978) and argillaceous deposits containing organic matter (Collins, 1967). Similar to bromide and chloride, iodine is very soluble (as iodide ion I^-) and does not readily substitute into mineral phases because of its large ionic size (Frape and Fritz, 1987). Although not particularly abundant, it is therefore concentrated preferentially in the aqueous phase during water–rock interactions which makes it a favorable tracer of salinity sources. Because of its relatively narrow range, Whittemore and Pollock (1979) consider iodide a potentially useful tracer (as I/Cl) to distinguish among different types of brines. Lloyd and others (1982) used (a) the ionic ratio of I/Sr to distinguish between different water sources in an alluvial aquifer in Peru, (b) I/Cl ratios to distinguish between old and modern saline ground waters in the Lincolnshire Chalk aquifer of England, and (c) I/Cl to distinguish saline water in a limestone aquifer from saline water in gypsiferous beds in Qatar. A very irregular distribution of iodine concentrations result from the application of fertilizers containing iodine (Lloyd and others, 1982).

The concentration of iodide in most fresh and brackish waters is less than 1 mg/L. Carbonate horizons in soil profiles appear to act as natural barriers to iodine migration (Whitehead, 1974). Iodide concentrations are also low in sea water (0.05 mg/L; I/CL = 2.6×10^{-6}), whereas oil-field waters may contain several tens or hundreds of milligrams per liter of iodide (I/Cl $>10^{-5}$). Collins (1969) reported unusually high concentrations of iodide in excess of 1,400 mg/L in some Anadarko Basin brines. Low I/Cl ratios in sea water and high I/Cl ratios in most oil-field and deep-basin brines may allow differentiation between these two sources of salt water.

Isotopes: The stable isotopic composition of a water sample may be indicative of the source of the water or of the source of the mineral content dissolved in the water. Carbon and sulfur isotopes reflect water–rock interactions and thereby may mask the origin of the water. At low temperatures, oxygen and hydrogen isotopes are much less affected by water–rock interaction than by fractionation processes before recharge to the water table and, therefore, are more indicative of the origin of the water than of the dissolved mineral content. Stable isotopes of oxygen and hydrogen (^{18}O and ^2H, deuterium) are used frequently to differentiate between waters originating at different recharge areas. This is based on the occurrence of fractionation processes between lighter and heavier isotopes, that is, (a) during evaporation, the heavier isotope becomes abundant in the solution and the lighter isotope becomes abundant in the vapor phase and (b) during precipitation, consecutive rainfall events become lighter in their isotopic composition. Therefore, coastal rain is isotopically heavier than rain occurring further inland. Also, evaporatively concentrated recharge water is isotopically enriched when compared to recharge water not affected by evaporation prior to infiltration. Enrichment or depletion is expressed relative to the

194

isotopic composition of standard mean ocean water (SMOW), arbitrarily being assigned a $\delta^{18}O$ and δD composition of zero. Within humid climates, the standard meteoric water line defined by Craig (1961) represents the relationship between oxygen-18 and deuterium in most recharge water. Arid zone recharge waters are better represented by a meteoric water line that is shifted toward isotopic enrichment (Welch and Preissler, 1986). Oxygen-18 and deuterium can be useful to distinguish local meteoric water that dissolved halite in the shallow subsurface from a regional oil-field brine (for example, Richter and Kreitler, 1986a,b) or any other brine that is derived from a different recharge area. In a study of formation waters from central Missouri, Banner and others (1989) used these ratios to trace the origin of the water all the way to the Front Range of Colorado. However, a basic problem exists in the fact that the water component of a brine may have a very different origin than the majority of the solute components, that is, water chemistry and isotopes do not necessarily reflect the same source (Kreitler and others, 1984, Frape and Fritz, 1987). At high temperatures, waters will equilibrate with oxygen-18 of the aquifer material and therefore will not preserve the original recharge signature. This oxygen shift can be of value in differentiating deep basinal waters from shallow meteoric waters.

Age-dating techniques using unstable isotopes such as carbon-14, tritium (^{3}H), or chlorine-36 may allow differentiation between old and modern sea-water intrusion (Lloyd and Howard, 1979, Custodio, 1987, Gascoyne and others, 1987). Carbon-14 and tritium are produced by cosmic rays interacting with nitrogen in the outer atmosphere. Because of this constant creation of carbon-14, the ratio of $^{14}C/^{12}C$ in the atmosphere is relatively constant. Studies of carbon-14 content in tree rings suggested that the $^{14}C/^{12}C$ ratio has varied only slightly during the past 7,000 years (Freeze and Cherry, 1979). However, when out of contact with the atmosphere, this ratio will decrease as a result of radioactive decay of the carbon-14 isotope, allowing determination of the time span that has elapsed since isolation from the atmosphere. This determination is complicated, however, by the addition of "dead carbon" from dissolution of carbonate rocks, as these rocks contribute carbon essentially devoid of carbon-14 (Freeze and Cherry, 1979). Large-scale atmospheric testing of thermonuclear weapons in the 1950's and 1960's created large amounts of tritium, which now enables differentiation of pre-1950's waters (0–2 tritium units) from post-1950's waters (>2-3 TU). Tritium has a half-life of only 12.3 years, which restricts age determinations to only a few tens of years. The half-life of carbon-14 of 5,730 years, in contrast, allows age determinations of organic matter or carbon dissolved in water in the order of tens of thousands of years (maximum of approximately 50,000 years). Chlorine-36 combines the advantages of the relatively conservative nature of the chloride ion with a half-life of approximately 300,000 years. As such, it may be more useful than tritium in the future, when the bomb-tritium peaks will have decayed as a result of its short half-life. Chlorine-36 is produced naturally by (1) spallation of heavier nuclei, such as argon, potassium, or calcium, by energetic cosmic rays, (2) slow neutron activation of argon-36, and (3) by neutron activation of chlorine-35 (Bentley and others, 1986). In addition, neutron activation of chlorine-35 in sea water by weapons testing in the South Pacific between 1952 and 1958 caused enrichment of chlorine-36 (Bentley and others, 1986). The difficulty of analyzing small quantities of the isotope in the past have been overcome in recent years by the use of Tandem Accelerator Mass Spectrometry. However, costs are high (up to $750 per sample) and existing data are sparse, both of which limit the wide-scale usefulness of this isotope in salinization studies at this time.

Strontium isotope ratios ($^{87}Sr/^{86}Sr$) may be useful to distinguish brines from different oil pools or stratigraphic units (Chaudhuri, 1978), reflecting strontium isotopic ratios of the host rocks. This isotope ratio is not used routinely in salinization studies and is probably best used to test a theory that was previously established using more conventional methods.

Sulfur isotopes ($^{34}S/^{32}S$) may be used to identify the source of sulfide or sulphate in ground water, which in turn may be useful to differentiate between salinization sources. For example, sulphate may originate from the solution of evaporite minerals, such as anhydrite or gypsum, giving the water a characteristic isotopic composition typical for the unit in which the minerals occur. Solution of sulfide minerals or decomposition of organic matter may provide a different isotopic composition that is characteristic and distinct for another water-bearing unit. In a study in Ohio, Breen and others (1985) suggested the use of sulfur isotopes to distinguish between isotopically heavy sulfate in brines and isotopically light sulfate in ground water. Dutton and others (1989) used this ratio to differentiate between two potential brine sources in parts of West Texas.

Determination of isotopes requires sophisticated laboratory equipment and techniques that are not routinely provided by just any laboratory. Certain isotopes are only determined in one or just a few laboratories in the country and costs can easily run into the hundreds of dollars per water sample. At a combined analysis cost of approximately $100 per sample (Table 13), determination of oxygen-18 and deuterium is probably too costly to be done on a routine basis but may prove very helpful for individual samples identified by other methods as being characteristic of a certain water type or water source. Sampling for oxygen-18 and deuterium is relatively easy, requiring only 250 or 500 mL of filtered water. Similarly straightforward is collecting water samples for tritium and sulfur-34. A one liter glass bottle is filled completely with filtered sample water for later analysis of tritium content. Laboratory analysis of sulfur-34 requires field treatment of the filtered sample, using small amounts of HCl and Cd-acetate or Zn-acetate. Much more elaborate is field-sample collection and preparation for carbon-14 determination (Feltz and Hanshaw, 1963; International Atomic Energy Agency, 1981). Depending on the amount of dissolved carbon in the water sample, conventional methods of carbon-14 analysis may require processing of hundreds of liters of water to precipitate the typically required 4 grams of carbon, using CO_2-free ammoniacal strontium-chloride solution as precipitation agent. For example, 2,000 liters of sample water would be needed at a bicarbonate content of 10 mg/L. However, the required amount of sample water would only be 40 liters at a bicarbonate content of 500 mg/L. At a cost of laboratory analysis in the order of $250 per sample, determination of carbon-14 may not be practical in most investigations. This is especially true for the even costlier method of determining carbon-14 using the Tandem Accelerator Mass

Spectrometry (cost at approximately $450 per sample), which otherwise has the big advantage of requiring much smaller amounts of carbon (in the milligram range), greatly reducing the amount of water needed.

In summary, collection and measurement of isotopes are expensive, often elaborate, and constitute techniques that may not be conclusive as tracers of salt-water sources by themselves. More often and probably more importantly, isotopes are used to support conclusions drawn from the evaluation of chemical constituents.

Lithium: Once in solution, lithium is not readily removed by exchange reactions or secondary minerals and it will accumulate in solution depending on time and availability in the host rock (Frape and Fritz, 1987). Therefore, it may be used as a good indicator of the degree of water–rock interactions. Whittemore and Pollack (1979) observed a wider range of Li/Cl ratios than of Na/Cl, Br/Cl, and I/Cl ratios in Kansas oil-field brines as a whole, but much narrower ranges in Li/Cl ratios than in the others when restricted to a particular geographic area. This would suggest that Li/Cl ratios may be good indicators for local salinization sources. However, lithium concentrations in most fresh and brackish ground waters are in the µg/L range and close to analytical detection limits. Therefore, lithium is probably used best at high chloride concentrations, that is, at low degrees of dilution of the original salinization source.

Magnesium: Magnesium in ground water is derived mainly from dolomite, limestone, and ferromagnesian minerals. In limestone terrain, an increase in magnesium concentration along the flow path is likely because dissolution of limestone causes an increase in magnesium but subsequent precipitation of calcium carbonate removes only little magnesium from solution (Hem, 1985). Dolomite solution and subsequent precipitation of calcite may also contribute to high magnesium concentrations, as suggested by Senger and others (1990) for the Glen Rose Formation of Central Texas. However, physical and chemical processes that may control the amount of magnesium are multiple, including dispersion, complexation, adsorption, desorption, precipitation, and solution. Concentrations of magnesium are less than calcium concentrations in most natural ground waters. In sea water, however, magnesium content is more than three times that of calcium (1,359 mg/L Mg vs. 410 mg/L Ca; Goldberg and others, 1971). Therefore, the Mg/Ca ratio may allow detection of sea-water intrusion into coastal fresh-water aquifers as long as ion exchange, dissolution, or precipitation does not affect either ion to a large degree. Because of its nonconservative nature, magnesium concentrations or Mg/X (X = Ca, Na, K, SO_4, Cl) ratios are used within a suite of other ratios but seldom constitute the sole indicator of a salt-water source. Like other major cations, determination of magnesium is included in standard ICP (Inductively Coupled Plasma–Atomic Emission Spectrometer) techniques at a relatively low cost.

Major Ions: Most chemical analyses of ground water include only the major cations Ca, Mg, Na, and possibly K, and the major anions HCO_3, SO_4, and Cl. In addition, many analyses reported in the literature, especially those being 30 years old or older, include a calculated value for Na+K, determined from the difference in meq/L of major anions and the cations Ca+Mg. This practice prevents a quality check (ion

197

balance) of the analytical work and should be avoided whenever possible. Analytical problems and costs are the main reason for the often small number of parameters reported. In salinization studies, major cations and anions are used mostly within concentration ratios related to chloride. Ratios are generally preferred to absolute concentrations because dilution of salt water by fresh water affects ratios of constituents to a smaller degree than individual concentrations. Chemical reactions, such as ion exchange, precipitation, and dissolution, change ionic ratios and present obstacles for the use of these ratios in determining salt-water sources at low concentrations. In many instances, chemical reactions between water and the aquifer material mask salt-water sources at low concentrations. The conservative ion chloride can sometimes be used to evaluate the degree of chemical reactions. For example, if the chloride concentration indicates a salt-water to fresh-water ratio of 1:10 in the mixture (calculated from known end-member concentrations) but calcium concentrations indicate a 1:5 mixture, the difference in calcium concentration may be explained by water–rock interactions (see also chapter 6 for calculation of mixing ratio). Therefore, chemical ratios of major cations and anions have to be applied carefully, keeping in mind chemical and physical processes that may change these ratios.

Sampling for major cations and anions (for procedures see Brown and others, 1970) requires two 500 mL, polyethylene bottles. Both samples are filtered using a 0.45 μ membrane filter. In addition, the cation sample is treated with a HCl solution to a pH below 3.0. Laboratory costs for all major cations and anions combined amount to approximately $60 per sample (Table 13).

Minor Ions: In contrast to the major ions which exist in soluble form in many different environments in the lithosphere and hydrosphere, minor elements and trace elements are less abundant and are often concentrated only under certain conditions. Concentrations seldom exceed 1 mg/L, but nevertheless these elements may reflect differences in aquifer compositions between different sites; differences that are not reflected in concentrations or concentration ratios of major ions. Anthropogenic factors may often lead to unnaturally high concentrations of minor constituents, creating excellent tracers of ground-water contamination. On the other hand, minor and trace elements in concentrated brines often occur in concentrations so small that significant dilution with fresher waters renders them close to or below detection limits. Also, many of the minor and trace elements may easily be precipitated or absorbed during flow and mixing with other waters due to a change in the chemical environment. For example, barium and strontium can be removed as sulfates when encountering water that has dissolved gypsum or anhydrite (Whittemore and Pollack, 1979). The occurrence of iron in one of its oxidation states, which govern its solubility, is very dependent on pH and Eh, making Fe an unfavorable tracer. In general, precipitation as oxides, hydroxides, carbonates, sulfates, phosphates, etcetera, removes many of the minor and trace elements from solution. Physical and chemical processes that may control amounts of these elements in ground water are dispersion, complexation, acid-based reactions, oxidation-reduction, precipitation-dissolution, and adsorption-desorption (UNESCO, 1980).

Nitrate: Excessive concentrations of nitrate in drinking water may cause methemoglobinemia ("blue baby" syndrome). Drinking water standard has been set at 44 mg/L nitrate, which is equivalent to 10 mg/L nitrogen. This standard is increasingly exceeded, especially in rural areas, where a variety of nitrate sources exist, and where private well-water use is still high. Sources of nitrate include (a) natural soil nitrogen converted to nitrate by bacteria, accelerated by cultivation of land, (b) septic tank drainages, (c) animal waste effluents, and (d) commercial fertilizers.

Nitrogen also occurs in ground water in the form of nitrite (NO_2^-) and ammonia (NH_4^+); concentrations of these compounds are generally much smaller than nitrate concentrations, however. Nitrate content has been used widely to identify pollution from organic waste, that is, from septic tanks and animal wastes, especially when combined with a relatively high chloride content, which is also present in those waters. In contrast, high chloride concentrations in association with low nitrate concentrations has often been used for identification of a salinization source other than agricultural effluents or septic tanks, as all other salinization sources (oil-field brine, halite, road salt, etcetera) are low in nitrate.

Where the source of nitrate is unknown, the isotopic composition ($^{15}N/^{14}N$) can be useful for differentiating soil nitrate from fertilizer nitrate or from organic-effluent nitrate (Kreitler and others, 1978; Kreitler, 1979). However, determination of nitrogen isotopes is relatively expensive (Table 13).

To prevent bacterial creation or destruction of nitrate between sampling and analysis, it is important to treat the water sample in the field. Thompson and Custer (1976) reported concentration increases in the order of 100 ppm in untreated samples relative to acidified samples within three days of sample collection. After several months, the same authors also detected small nitrate increases in acidified samples. Therefore, Thompson and Custer (1976) recommend that nitrate water samples should be filtered, acidified, cooled, and analyzed quickly. R. F. Spalding (personal communication, 1991) recommends filtering and freezing of the samples without acidification before shipment to the laboratory in frozen form (note that sample bottles should not be filled completely to avoid breakage).

Potassium: Potassium is added to ground water mainly from the weathering of feldspars and clay minerals. But more easily than it is incorporated into solution it is taken out of solution by fixation into clay minerals (Craig, 1970, Frape and Fritz, 1987). Leaching of potassium from illite is part of the salinization process observed at saline seeps in Montana (Donovan and others, 1981), suggesting its potential use as a salinity tracer. In Ohio, Breen and others (1985) demonstrated that the ratio of K/Na can be useful to distinguish between brines from different sandstone units. High concentrations are often found in mine waters, hot-spring waters, and in sylvite-dissolution brines. Potassium concentrations are often lowest in halite dissolution brines, highest in brines from carbonate aquifers, and intermediate in brines from sandstone units (Rittenhouse and others, 1969). Despite these differences, however, potassium concentrations or K/X ratios (X = Ca, Mg, SO_4, Cl, Br) are mostly used within a suite of other constituent ratios for detection of salinization sources.

Selenium: Selenium in ground water is a byproduct of pyrite oxidation and dissolution and is often found in high concentrations in irrigation and saline-seep areas. Areas underlain by black shales or shale-derived materials (Donovan and others, 1981; Deverel and Gallanthine, 1989) appear to be affected most. Selenium minerals are also associated with some of the uranium deposits in sandstones in the Western United States. As such, its usefulness as a salinization tracer may be restricted to certain geographic areas. Recommended limit of selenium in drinking water is 0.01 mg/L (U.S. Environmental Protection Agency, 1975)

Silica: With the exception of some high-temperature waters, silica concentrations in most natural waters are commonly less than 100 mg/L. Fogg and Kreitler (1982) observed a general decrease in silica content in ground water away from recharge areas, which could be explained by precipitation of quartz cement or precipitation in authigenic clays. In oil-field brines, silica concentrations are highly variable (Rittenhouse and others, 1969), depending largely on lithology, residence time, and temperature. Silica content has been used in a study in the Netherlands to distinguish young salt water from old salt water (Custodio, 1987), but little other work related to salt-water sources has been documented.

Sodium: Sodium in ground water is derived mainly from the decomposition of feldspars and from sodium salts. Once in solution, it tends to stay there unless minerals having high cation-exchange capacities and available exchange sites are present. Sodium is commonly used in salt-water studies as a constituent ratio of Na/Cl, with a weight ratio of 0.65 in brine being characteristic of halite dissolution and a ratio of less than 0.60 in brine being characteristic of oil-field/deep-basin brine (for example, Leonard and Ward, 1962; Oklahoma Water Resources Board, 1975; Whittemore and Pollack, 1979; Gogel, 1981; Richter and Kreitler, 1986a,b). Cation exchange between Ca in the aquifer material and Na in the water leads to a decrease in Na/Cl ratios. Albitization of plagioclase or K-feldspar may also lead to a decrease in Na/Cl ratios (Land and Prezbindowski, 1981). This is the case in most oil-field/deep-basin brines and wherever Na-type water replaces Ca-type water in the presence of suitable exchange sites. An increase in Na/Cl ratios will occur when fresh water replaces marine water (Custodio, 1987). The Na/Cl ratio is applied most successfully to differentiate halite-dissolution brine from oil-field/deep-basin brine, as documented in numerous studies in Kansas, Oklahoma, and Texas and to document mixing of fresh water with salt water. It usually is not applied to differentiate between brines from different stratigraphic units because of a general overlap in ratios. In combination with other constituents, however, sodium may be used successfully to distinguish between different brines on a local basis. This was shown by Breen and others (1985), who used the ratio of K/Na to distinguish between brines from three sandstone units in eastern Ohio.

Sodium makes up a major portion of the cation composition in most ground waters, but often is determined by the difference between the sum of anions and the sum of calcium plus magnesium, expressed as meq/L. This practice should be avoided whenever possible, because analytical errors that could be detected through cation-anion balance of analyzed constituents will remain undetected.

200

Sulfate: Sulfate occurs in often high concentrations in nonsaline ground water where it is derived primarily from the decomposition of iron sulfides and solution of gypsum. To a lesser degree it is derived from volcanic sources, as a result of the combustion of fossil fuels, and from industrial and mining activities (Craig, 1970). Sulfate is easily reduced bacterially below about 80°C and thermally at higher temperatures; therefore, sulfate-dominated solutions are rare in the deep subsurface (Land, 1987). Because of these low concentrations in most brines, the constituent ratios of SO_4/Cl or SO_4/TDS have been used occasionally for detection of brine mixing with fresh water. However, because sulfate concentration can be readily altered by chemical and biochemical processes, the usefulness of this parameter in salt-water studies is limited. Therefore, ratios of SO_4/X (X = Cl, TDS, Ca+Mg) are used best within a suite of ionic ratios. The recommended maximum limit of sulfate in drinking water is 250 mg/L (U.S. Environmental Protection Agency, 1975).

Total Dissolved Solids: The concentration of TDS in ground water is an overall measure of water quality. Certain ranges of TDS values are often used to define terms such as fresh, brackish, saline, or brine. Many different classification ranges are being used in the literature to define the same or similar terms; the two major ones are shown in Table 2.

The amount of TDS is a general indicator of water quality and is a useful parameter for quality monitoring. In studies of saline water, TDS is closely related to Cl concentrations and is used primarily in mapping to illustrate the location and extent of poor-quality ground water. TDS is of limited use in identification of salt-water sources except in simple one-source scenarios.

The concentration of TDS in solution is determined either as the dry residue after evaporation or is calculated from the individual concentrations of major cations and anions. For calculation of TDS, HCO_3 concentrations are converted to carbonate in the solid phase using a gravimetric factor (mg/L HCO_3 * 0.4917 = mg/L CO_3); this assumes that half the bicarbonate is volatilized as CO_2 and H_2O and that the carbonate value obtained corresponds to conditions that would exist in dry residue (Hem, 1985).

4.2. Summary of Field Techniques

The field-sampling methods described next follow for the most part the generally accepted procedures, and were applied successfully by the authors in numerous salt-water studies. Some of the suggestions made here may not meet with everyone's approval. For further references on sampling methods, the reader may refer to Brown and others (1970), Wood (1976), Lico and others (1982), and Hem (1985).

Collection of ground-water samples within salinization studies will be either from established wells, from testholes drilled for the particular study, or from surface waters. The purpose of ground-water sampling is to collect a water sample representative of the aquifer of interest. Installation of monitoring wells and sample collection induce changes, which may make it impossible to obtain a truly representative sample (Pennino, 1988). Fortunately, these changes are most profound on dissolved gases, organics, and trace metals, which are of minor importance in most salt-water studies.

Sample recovery will generally be with the help of a pump or a bailer. Wells may be pumped or bailed for some time before sampling to avoid collection of a water that has been changed through long storage time in the well bore and is not representative of true formation water. Pumping or bailing of three bore volumes before sampling is generally considered sufficient. Garner (1988) recommended that sample collection may be started when in-line monitoring values of pH, Eh, temperature, electrical conductivity, and dissolved oxygen do not vary more than 10 percent per casing volume pumped. Puls and others (1990) suggested that turbidity be monitored in addition to these parameters, and that the pumping rate should be close to the actual ground-water flow rate. Some constituents, such as lead and cadmium, however, may not stabilize even if those monitored water-quality parameters have stabilized, as reported by Pennino (1988).

Sampling for most constituents will require filtering in the field (0.45 μm filter), which is done very easily in the case of pumped samples using in-line filters. Filters will be discarded after each sample to avoid cross-contamination between samples, which, at a cost of approximately $15.00 per filter may appear expensive at first but probably is the most convenient, time-saving, and safest sampling method. Although filtering in the field is a standard technique, Snow and others (1990), who compared analytical results of filtered and unfiltered water samples, did not find a statistical difference between these two groups. When planning sampling of salt-water wells it may be an advantage to conduct sampling from the location of lowest salinity to the location of highest salinity, which makes equipment cleaning between sampling easier and decreases the probability of contaminating a low-TDS water with a high-TDS water through the repeated use of sampling equipment. Repeated use of sampling equipment requires generous rinsing of equipment using deionized water. A convenient check that sufficient rinsing was performed is provided by addition of some drops of silver nitrate into a sample of rinse water. In the case of insufficient rinsing, the sample will turn cloudy from the formation of silver chloride, indicating that cleaning should be continued. Such a check is needed especially when bailing samples, because those need to be filtered using a specially designed reusable filter chamber, into which disposable filter paper is inserted before the sample is forced into sample bottles by nitrogen gas. The use of nitrogen gas is recommended when contact with atmospheric gas is to be kept at a minimum. In most salinization studies, however, atmospheric contact is of minor concern and a foot pump may suffice for forcing the samples through the filter.

Special care has to be taken when collecting water samples from oil and gas wells. To avoid too much contact of sample equipment with oil, oil—water mixtures are collected in a bucket with a drum tap at the bottom. After several minutes, water and oil will separate and water can be drained from the bucket. Draining through a glass wool filter will remove most of the residual oil in the water before transfer to the filter bottle for final sample collection. Afterwards, the filter bottle may require cleaning with soap and plenty of deionized water.

In addition to samples collected for laboratory analyses, some relative unstable parameters are generally determined in the field. These include pH and alkalinity, which are determined during and through titration. It is general practice to perform this alkalinity titration as soon as possible after sampling to avoid out-gassing of the sample. In most salinization studies, the alkalinity value is used for not more than the cation–anion balance or for the plotting of ionic percentages in Piper diagrams, in which cases slight changes are of little concern. Therefore, collection of a separate sample for alkalinity titration to be performed at the end of a sampling day can greatly increase sampling efficiency without adverse effects on sample quality and data evaluation. It is the authors' personal experience that alkalinity titration performed on a completely filled sample bottle within 12 hours of sample collection has produced satisfactory sample results, that is, cation-anion balances always satisfied a pre-determined error tolerance of less than three percent.

It is a good practice to design field-data sheets in the office before going into the field. These data sheets will include a form that specifies the samples to be taken and the bottles and preserving agents to be used. This information can be included on a sample-summary sheet, to be filled out at each sample point. A different form will be used for the alkalinity titration.

5. DATA AVAILABILITY AND SELECTION

5.1. Sources of Data

Among the variety of available data sets, the one maintained by the U.S. Environmental Protection Agency, STORET, is probably the largest. Other data sets are maintained by the U.S. Geological Survey, WATSTORE, by individual state agencies (for example, Texas Natural Resources Information System [TNRIS], in Texas), and by commercial data services (for example, Petroleum Information, Dwights). These large data bases are available on magnetic tape, on diskette, or as printouts either for a fee or free of charge. Oftentimes, site-specific data can be retrieved from these sources. Local water-chemistry laboratories, individual researchers, and published reports can be excellent sources of data not included on some of the federal, state, or commercial data banks. However, these data are mostly available only as paper copies. Sorting and compilation of chemical analyses can be very costly, as indicated by Hiss (1970), who reported acquisition and data reduction costs of approximately $5.00 per usable record during a study of saline water data from southeastern New Mexico and western Texas. These costs appear relatively high but are much less than costs of field collection and laboratory analyses, which may run into several hundreds of dollars per analysis.

Quality and completeness of analyses vary to a large degree between and within data sets. Some of the problems encountered when working with existing data sets are described below, using the following data retrievals conducted for this study. Retrieval #1 (STORET): Complete chemical analyses from water wells and springs in the United States in which chloride concentrations are greater than or equal to the drinking water standard of 250 mg/L. This request resulted in 99,915 analyses from 18,772 stations. Retrieval #2 (STORET): Locations and chloride concentrations of water wells in the United States. This request resulted in 734,091 chloride analyses from 212,678 stations. Retrieval #3 (TNRIS): Locations and chloride concentrations of all water wells in Texas. This request resulted in 71,835 chloride analyses from 33,463 stations.

The completeness of analyses varied to a high degree in Retrieval #1, that is, only approximately 50 percent of all analyses include all major cations and anions. Selection of minor constituents or isotopes reduces the availability, and with it the use of this data set, even further, as, for example, only 2,716 bromide values, 1,466 iodide values, and 826 oxygen-18 values are included in the set. Requesting complete analyses or a large number of parameters per analysis, in this case resulted in a huge data file that is difficult and expensive to work with. Separation into several small data files, for example, one of major constituents and another one of isotopes, will eliminate some of the problems of working with large files.

Although a relatively good coverage for almost the entire country was provided by Retrieval #2 (Fig. 71), it is apparent that there is a lack of data in some areas, such as Texas. This lack is artificial, as indicated by Retrieval #3 (Fig. 72), which illustrates a much higher data density from the TNRIS data set when compared to the STORET data set. This suggests that working with more than one established data set can greatly increase data coverage and that care should be exercised during selection of a certain data bank.

Existing data sets often contain a bias toward certain data. For example, an evaluation of ground-water chlorinity in the state of Tennessee using the STORET data base would actually be an evaluation of water quality from a relatively small area in Tennessee (Fig. 71). Obviously, most of the samples available for this state were derived from a study that was confined to a certain area. Site-specific studies also may tend to be biased toward a specific source of water, such as shallow ground water of potable quality (biased toward fresh water), sources of ground-water contamination (biased toward saline water), ground-water quality in formation A (biased toward fresh or saline water), or ground-water quality in a specific county. Most large, computerized data bases of water quality (with the exception of petroleum-related data banks) consist of analyses from observations wells and from municipal and domestic wells, which by their own nature are for the most part of the best quality available in any area. This may be reflected in Retrieval #2; 88 percent of the stations have chloride concentrations in water that are less than the drinking-water standard of 250 mg/L. Whenever a poor-quality water is encountered during drilling or when a water well has gone bad, the well is most likely abandoned, no additional analyses are obtained, and a new well furnishing better water is installed. Indiscriminant mapping may identify this as an improvement in water quality in the area or it may be interpreted as an area without salinization problems. The small number of chloride analyses with chloride concentrations greater than 250 mg/L (9 percent of all stations with chloride analysis) suggests that the STORET data set may be biased toward low-TDS waters. Another problem with the use of of large, existing data sets is the often unknown source of the data and the lack of information regarding well depths or producing formations. For example, only approximately 30 percent of all analyses in Retrieval #1 included some kind of depth specification. A depth specification is important, however, as most of the country is underlain by saline ground water at some variable depth.

Besides data bases of chemical analyses, literature data bases can be used to assist in data searches. For example, for this report, the Water Resources Abstract data base, which contains approximately 170,000 records on diskette, was used for screening of keywords, such as salt water, brine, salinization, saline seep, etcetera. A variety of computer-based data bases are available to researchers (Table 14), as described by Atkinson and others (1986) and Canter (1987), for example.

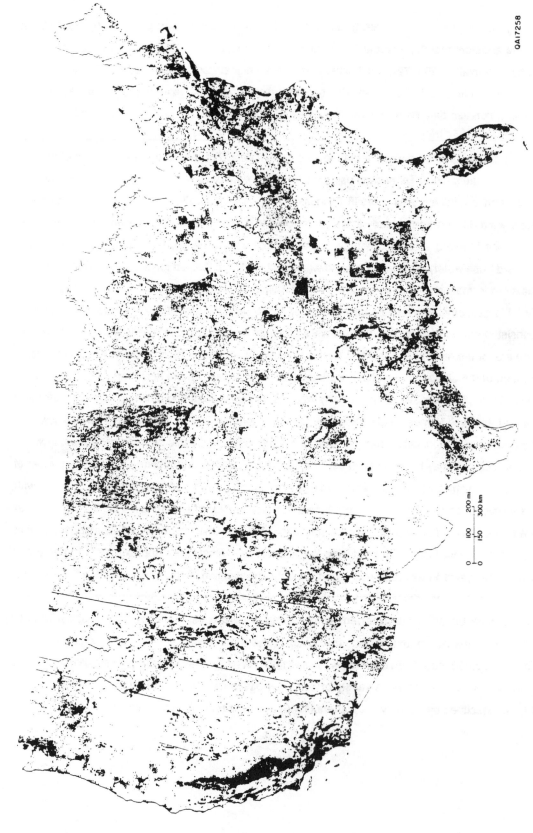

Figure 71. Location map of ground-water stations for which a chloride value is available at U.S. Environmental Protection Agency's data base STORET (Retrieval #2; approximately 200,000 stations).

QAI7258

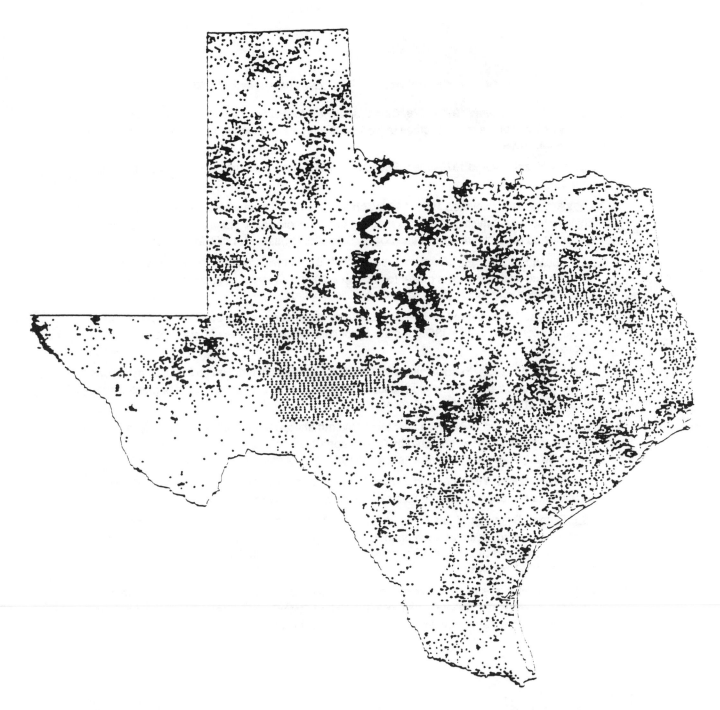

Figure 72. Location map of water wells in Texas for which a chloride value is stored at the Texas Natural Resources Information System data bank (approximately 33,000 stations).

Table 14. Description of literature data bases (from Canter, 1987).

1. Agricola. Years covered 1970–1978. Contains all the information of the National Agricultural Library and comprehensive coverage of worldwide journals and monographic literature on agricultural subjects.

2. Agricola. Years covered 1979–1984. See above. The two Agricola data bases contain 2 million records.

3. Compendex. Years covered 1970–1984. Contains the machine-readable version of the Engineering Index which includes engineering and technological literature from 3,500 journals and selected government reports and books. Size of data base—1.4 million records.

4. Conference Papers Index. Years covered 1973–1984. Contains records of more than 100,000 scientific and technical papers presented at over 1,000 major regional, national, and international meetings each year. Size of data base—1.0 million records.

5. CRIS/USDA. Years covered 1982–1984. Contains active and recently completed agricultural research sponsored by the USDA or state agriculture institutions. Size of data base—31,000 records.

6. Dissertation Abstracts. Years covered 1961–1984. Contains dissertations from U.S. institutions as well as Canada and some foreign schools. Most of the abstracts are for degrees granted after 1980. Size of data base—852,000 records.

7. NTIS. Years covered 1964–1984. Contains government sponsored research, development, and engineering plus analyses prepared by federal agencies, their contractors and grantees. Unclassified, publicly available reports are available. Size of data base—1.1 million records.

8. Pollution Abstracts. Years covered 1970–1984. Contains environmentally related literature on pollution, its sources and control. Size of data base—107,000 records.

9. SSIE Current Research. Years covered 1978–1982. Contains reports of both government and privately funded scientific research projects either in progress or recently completed in all fields of basic and applied research in the life, physical, social, and engineering sciences. Size of data base—439,000 records.

10. Water Resources Abstracts. Years covered 1968–1984. Contains materials collected from over 50 water research centers and institutes in the United States and focuses on water planning, the water cycle, and water quality. Size of data base—173,000 records.

5.2. Selection of Data Criteria

Before starting with the evaluation of a set of existing chemical data, it is worthwhile to establish certain criteria that must be met by the data. These criteria will vary with the nature of information required. Exact well locations (for example, latitude and longitude) will be important in a study of a contaminant plume from a leaky salt-water well, whereas specification of a county code may be sufficient in a regional water-quality study. In the former case, analyses of waters from unknown locations will be discarded from the beginning or an effort will have to be made to identify well locations. Similarly, the date, method, or purpose of sample collection may be of great importance. The date allows identification of chemical changes over time and the establishment of precontamination concentrations. Mixing of current and historical data without dates, in contrast, may lead to wrong conclusions regarding the geographic distribution of water-quality changes. The method of sampling will determine how reliable analyses of certain constituents may be. For example, lab-determined alkalinity values may suffice in a study in which alkalinity values are used for not much more than a mass balance, but may not be acceptable in a study of equilibrium conditions within carbonate systems. The purpose of sampling may be reflected in an underlying bias toward a certain water type; for example, monitoring of municipal wells will be biased toward good-quality waters, whereas sampling within a salt-water contamination study will be biased toward poor-quality water. When using large, existing data bases, some of these parameters may not be available, especially when different sources contribute to that data base.

The problem of completeness of information includes the availability of certain chemical parameters. Most analyses of ground water include the major cations (Ca, Mg, and Na) and the major anions (HCO_3, SO_4, and Cl). Provided these constituents are all determined in the laboratory or in the field, a mass balance error can be determined (difference of sum of cations and sum of anions over the sum of cations and anions [in meq/L]; $|\Sigma cations - \Sigma anions|/[\Sigma cations + \Sigma anions]$). The number of data available, the quality of the water, and the personal quality criteria established by the investigator will determine what kind of error is considered acceptable. This error boundary will most likely be high when any sample discarded would represent a substantial loss, that is, when data coverage is very sparse, and when dealing with very fresh waters. The upper margin of error generally is located somewhere between 5 and 10 percent. In the case that one parameter, sodium, was calculated by difference instead of analyzed, the balance error will be zero percent. Availability of an adequate number of analyses that had not been determined in such fashion will conclude if calculated values should or should not be used.

The availability of existing data often determines the technique to be used for evaluation of water chemistry. For example, chloride and bromide concentrations are often used with good results to distinguish between halite-solution brine and oil-field brine. But because bromide is not a part of standard water analyses, only a relatively limited amount of data may be available. Also, time and money will often determine which technique or data base to use. Availability of a free data base (computer tape) of water chemistry from state or federal agencies can considerably cut down on costs and time, as opposed to having to compile data from published sources, water-quality files, or chemical laboratories. On the other hand, collection and analysis for isotopes may be very time consuming and expensive, as costs per single analysis may go into the hundreds of dollars (Table 13). When using commercial, outside laboratories for sample analyses, the following observations should be kept in mind (Rice and others, 1988): (a) the reliability of laboratory analyses should not be taken for granted, (b) analytical reliability may not be reflected in the price charged by laboratories, and (c) quality assurance programs benefit both the customer and the laboratory.

How to graphically display and statistically evaluate data selected from available sources or collected in the field will be discussed briefly in chapter 6.

6. GRAPHICAL AND STATISTICAL TECHNIQUES

Evaluation of chemical analyses often starts with graphical display and statistical manipulation of physical and chemical data. Which technique is used depends largely on the amount of data and on the type of information that is needed. In salt-water studies, techniques are used that maximize the separation of chemical characteristics between potential salt-water sources and illustrate to which salt-water source a contaminated water sample belongs.

6.1. Graphical Techniques

Graphical techniques are used to (a) illustrate the chemical character of a single analysis, (b) compare the characteristics of several analyses, (c) assist in identifying the relationship that exists between water samples, and/or (d) to calculate mixing ratios between fresh water and the contaminating source.

Among these techniques, the simplest ones illustrate a single parameter or analysis. Contouring or posting of chemical parameters onto maps or cross sections (Fig. 73) is done for locating areas of abnormal chemical composition, suggesting an erroneous value, a possible point source of mixing (for example, a leaky well), or more than one source of water. As such, these maps are also useful in identifying local positions of contaminant plumes or regional changes in water quality. In order to represent several ions that make up a chemical analysis either several contours, several maps, or other methods that combine the ions in a convenient form, need to be used. These other methods include bar graphs (Fig. 74a), pie charts (Fig. 74b), and polygonal (Stiff) diagrams (Fig. 74c).

On bar graphs and pie charts, concentrations of major cations and anions are illustrated by different sizes of representative areas. The overall shape and size of the graph may stay the same, in which case relative concentrations (percentages) are indicated. On bar graphs, absolute concentrations of individual ions can be displayed by changing the sizes of representative areas. A change in overall size of the pie chart, in contrast, can be used to express changes in TDS concentrations. Extending relative areas according to concentration scale similar to the one used in bar graphs can be used to present percentages as well as absolute concentrations on pie charts. Bar graphs were used by Williams and Bayne (1946) to display differences in mixtures of fresh water and saline-formation water from mixtures of fresh water with oil-field brine in Kansas. Relatively low percentages of magnesium and sulfate in oil-field waters are reflected in mixtures between oil-field brines and fresh water; mixtures of fresh water with saline formation water, in contrast, is characterized by higher magnesium and sulfate percentages. Pie charts are often used to illustrate water quality on maps, as was done on a national scale by Feth and others (1965).

Stiff diagrams allow a quick comparison between samples primarily through a change in shape, which is caused by different ionic compositions. This is provided by consistent plotting of cations to one side and anions to the other side of a vertical zero line and by connecting the end points. The distance of each

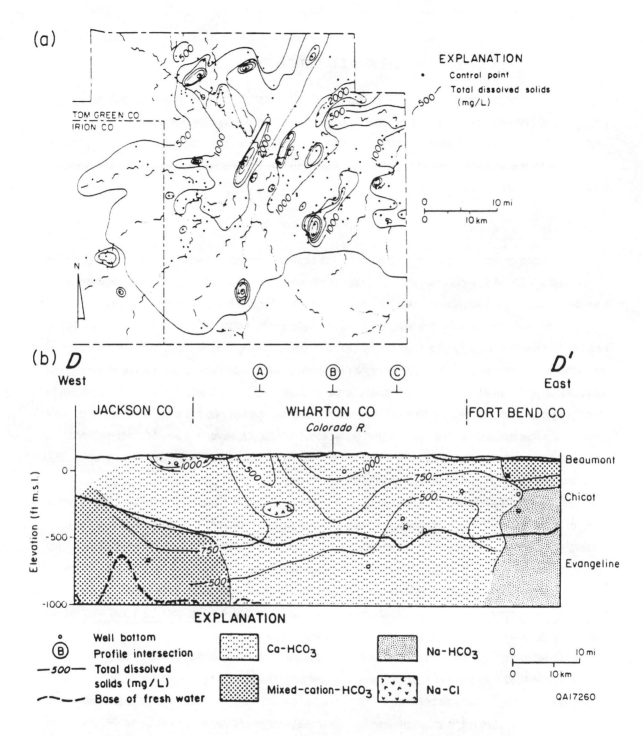

Figure 73. Graphical illustration of chemical analyses by contouring of individual parameters onto (a) maps and (b) cross sections (from Richter and others, 1990; and Dutton and Richter, 1990).

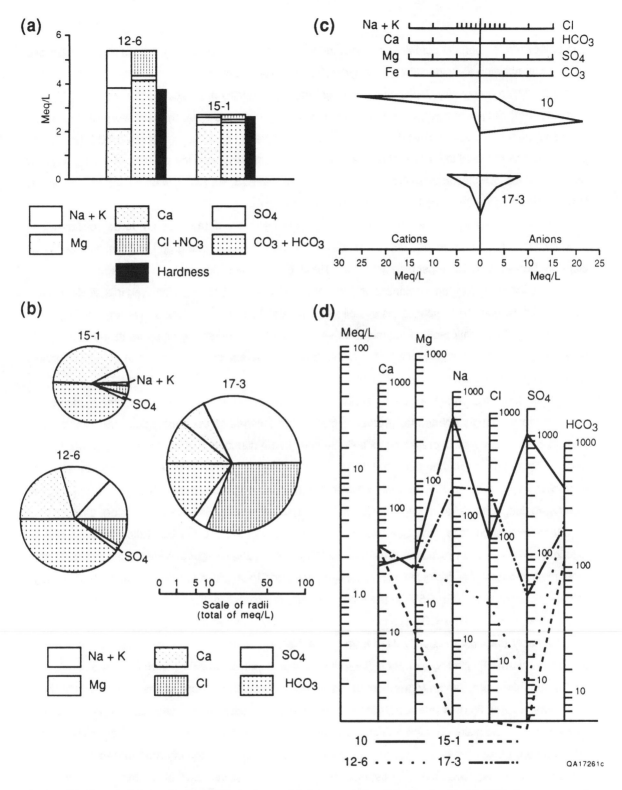

Figure 74. Presentation of major ions in form of (a) bar graphs, (b) pie charts, (c) Stiff diagrams, and (d) Schoeller diagrams (from Hem, 1985).

end point from the vertical zero line is proportionate to the concentrations (in meq/L) of the respective constituent, allowing easy identification of hydrochemical facies. Stiff diagrams were used by Burnitt and others (1963) to graphically depict changes in ground water quality in Ogallala outcrop areas of Texas (Fig. 51). Stiff diagrams and geographic mapping of major chemical parameters were also used by Levings (1984) in a study of oil-field pollution in the East Poplar field, Montana. Elevated TDS, Na, and Cl concentrations reflected the extent of ground-water movement away from the pollution site on isocontour maps, whereas Stiff diagrams illustrated the change from low-TDS, Na-HCO_3 waters (background levels) to high-TDS, Na-Cl waters (produced oil-field waters).

These single-analysis diagrams are of limited use when large data bases are to be considered. In such a case, a large number of analyses needs to be combined within one graph, such as in a Schoeller diagram (Schoeller, 1935), Piper diagram (Piper, 1944), or a bivariate plot.

On a Schoeller diagram, similarities and dissimilarities of water types are displayed in a similar fashion as in the Stiff diagram by comparing the overall shape created by all major anions and cations making up an analysis (Fig. 74d). This plotting technique is very useful for representing changes in the relationship between ions, as nonchanging ions plot at identical points whereas changing ions plot at different points along their respective axes.

Piper diagrams combine major cations and major anions in separate triangles, reducing the compositions to single points that represent meq/L percentages of each individual cation and anion (Fig. 75a). Each analysis is presented as a single point in the diamond-shaped diagram by projecting the anion and cation positions into that field. This technique allows convenient determination of hydrochemical facies (Fig. 75b), as well as assistance in interpreting relationships that may exist among the water samples, such as (a) mixing of different water types (see also Sea-Water Intrusion, chapter 3.3), (b) cation exchange (see also Sea-Water Intrusion, chapter 3.3), (c) precipitation and dissolution reactions, and (d) sulfate reduction (Custodio, 1987). For example, Krieger and Hendrickson (1960) used Piper plots to graphically depict the mixing between brine and fresh water, which was suggested from high chloride concentrations (Fig. 75). On the Piper diagram, the contamination is indicated by a straight-line mixing trend from a Ca-Mg-HCO_3 type fresh water to a Na-Cl type salt water.

Probably the most frequently used form of graphical presentation of chemical data is the bivariate plot (scattergram). By plotting a physical (for example, well depth, distance) or chemical (for example, Ca, Na/Cl) variable against another, the correlation between these parameters can be identified. On these plots, large scatter signifies little correlation between the plotted parameters, whereas any trend indicates either mixing of two waters or evolution of one water, such as, evaporation, precipitation, or solution. More than one large scatter may suggest two or more distinct groups of water whereas one or a few points outside a suspected trend may suggest erroneous data points or inclusion of nonrepresentative water samples resulting from, for example, a different origin (aquifer, well depth) or possibly point-source contamination. Any combination of parameters may be used, but it is preferred to plot parameters for

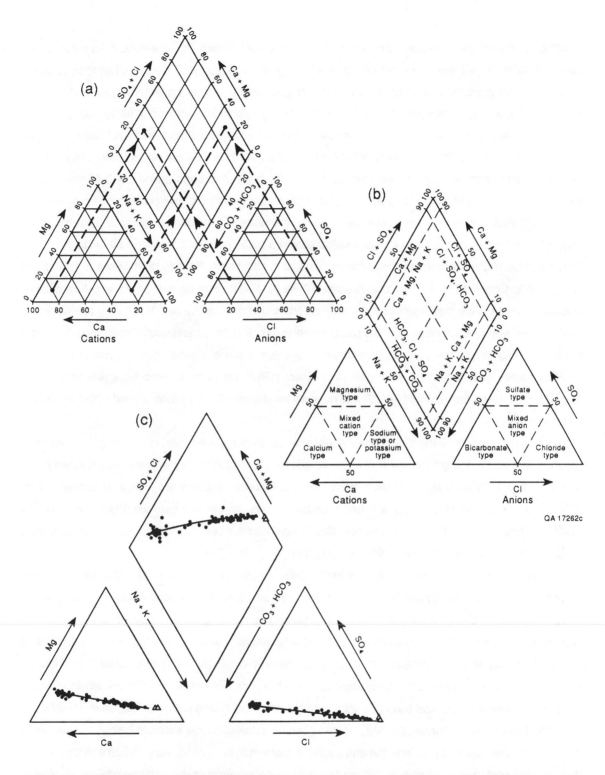

Figure 75. Presentation of chemical constituents on trilinear (Piper) diagram (a) (c) and classification scheme of hydrochemical facies (b), based on major-ion percentages (from Krieger and Hendrickson, 1960, and Freeze and Cherry, 1979).

215

which a correlation can be explained by chemical or physical processes. For example, a good correlation between calcium and sulfate or calcium and alkalinity strongly suggests chemical control of these parameters by gypsum and calcium carbonate, respectively. Another parameter pair that is used frequently, especially in salt-water studies, is that of sodium and chloride. Solution of halite (NaCl) by fresh ground water results in a saline water in which the molar concentrations of sodium and chloride are equal (mNa/mCl = 1) as long as the water is not too diluted (Na/Cl ratio is typically greater than unity in fresh water). Bivariate plots of sodium over chloride reflect this process through a slope of 0.65 on a plot of weight ratios (Fig. 76) or a slope of unity on a plot of molar concentrations. Oil-field brines also often show a good correlation between sodium and chloride, but the slope is considerably lower in most instances (Fig. 76). This difference in slope of Na/Cl ratios between halite-solution brines and oil-field/deep-basin brines has been used extensively for salt-water studies in Oklahoma, Kansas, and Texas (Leonard and Ward, 1962; Gogel, 1981; Richter and Kreitler, 1986a,b). Because of its conservative nature once in solution, chloride is the most often used parameter in this kind of bivariate plot. Regarding salinization of fresh ground water, representation of the composition of one or more potential salinization sources and a possible mixing trend between fresh ground water and one of these sources is of special interest. This technique was used successfully by Richter and others (1990), for example, who were able to separate one formation brine from another (Fig. 77) and disposal brine from naturally saline ground water (Fig. 49) in a salt-water study in West Texas.

When dealing with salinization of ground water which, in many instances, is a mixing between fresh ground water and saline ground water or brine, absolute concentrations vary widely but constituent ratios only vary to a small degree. In addition, because all salinization sources are high in dissolved solids, differences in constituent ratios are often better tracers of certain sources than are absolute concentrations. This led Novak and Eckstein (1988) to propose the use of ratios instead of concentrations in some of the traditional graphical techniques (Fig. 78).

When the end-member composition of fresh water and the contaminating salt water are known, the percentage of each endmember in a mixing water can be determined mathematically or graphically (Custodio, 1987). Using the most conservative dissolved ion in ground water, chloride (see also chapter 4), the percentage is calculated by the following equation or read from the percentage scale of figure 79, based on the assumption that mixing is the only dominant effect that caused the chloride increase in the mixing water. Chloride concentrations of the endmembers are used in the equation and in the graph, where they are end point of a mixing trend. After the mixing ratio has has been established, theoretical mixing values for Na, Ca, Mg, etcetera, can be calculated (see example below). Deviations of these theoretical mixing values from the true values measured in the mixing water indicate changes other than mixing, such as ion exchange. Other conservative constituents that can be used instead of or in combination with chloride are bromide (Fig. 79) and oxygen-18 (Arad and others, 1975).

216

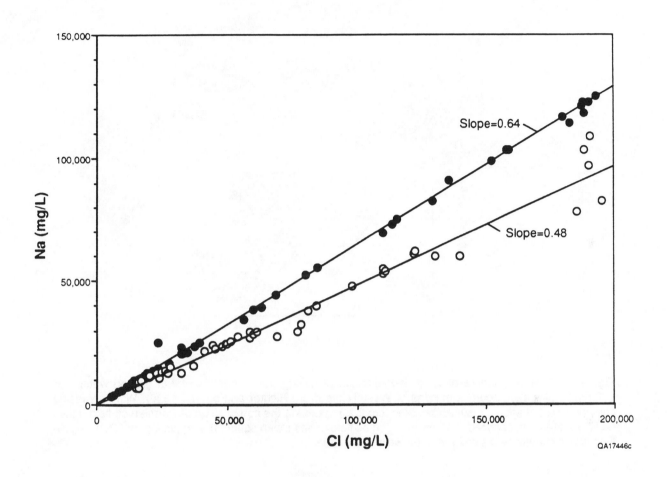

Figure 76. Bivariate plots of Na versus Cl for halite-solution (solid dots) and deep-basin brines (open dots).

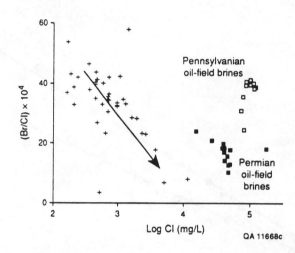

QA 11668c

Figure 77. Use of bivariate plots for identification of mixing trends between fresh ground water and potential salinization sources in parts of West Texas (from Richter and others, 1990). With increasing chloride concentrations, testhole samples (crosses) approach the composition of Permian oil-field brines (solid squares), as opposed to Pennsylvanian oil-field brines (open squares), suggesting that Permian oil-field brines contribute to salinity in water wells of the area.

218

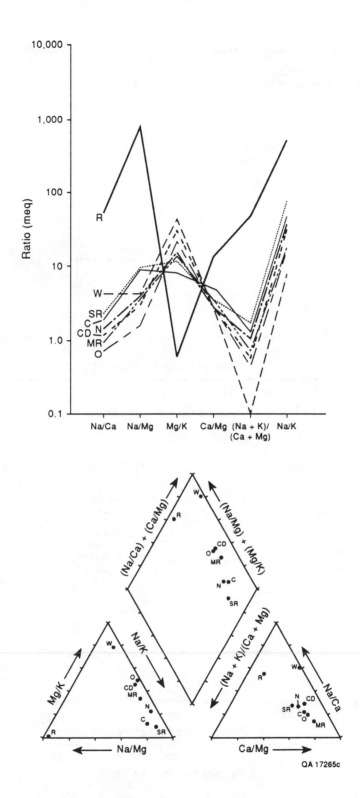

Figure 78. Modified Schoeller and Piper diagrams using concentration ratios as endpoints (from Novak and Eckstein, 1988).

219

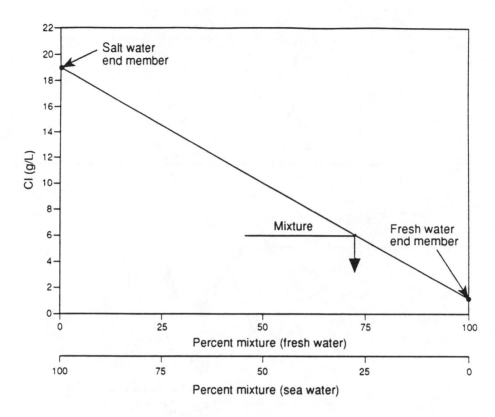

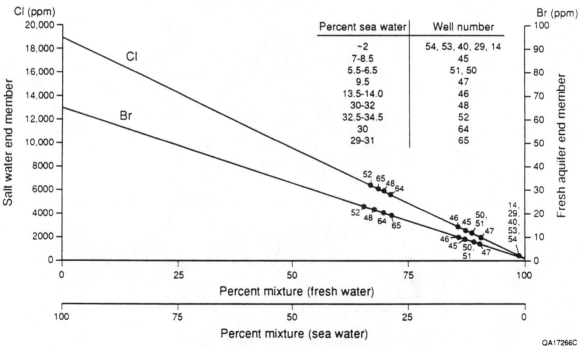

Figure 79. Calculation of mixing percentages between fresh water and salt water using mixing graphs of chloride and bromide (from Arad and others, 1975, and Custodio, 1987).

220

$$C_M = X \cdot C_F + (1-X) \cdot C_S ===> X = (C_S-C_M)/(C_S-C_F)$$

where:

C_M = Constituent concentration in mixing water,

C_F = Constituent concentration in fresh water,

C_S = Constituent concentration in salt water,

X = Fraction of fresh water in the mixture, and

1–X = Fraction of salt water in the mixture.

Example:

Step 1: Cl_M = 1,500mg/L; Cl_F = 50mg/L; Cl_S = 35,000mg/L

===> X = (35,000-1,500)/(35,000-50) = .96

The fresh-water source is represented by 96 percent and the salt-water source is represented by 4 percent.

Step 2: Na_M = 600mg/L; Na_F = 155mg/L; Na_S = 22,000mg/L

with 0.96 = (22,000–Y)/(22,000–155)

===> Y = 22,000 – 0.96*(22,000–155) = 1,029

The theoretical sodium concentration in the mixing water is 1,029 mg/L; the actual concentration is only 600 mg/L, however. Ion exchange may account for the loss of sodium, as long as another cation shows a comparable gain to its theoretical value.

6.2. Statistical Techniques

The application of statistics depends to a high degree on the number of observations in the data base and the nature of the required information. For the purpose of this report, only a general discussion of readily available statistical techniques was attempted. The exception is made with Stepwise Discriminant Analysis, which is presented as one possible technique to identify useful parameters for identification of salinity sources.

Hem (1985) pointed out that the literature abounds with questionable applications of statistical procedures. The major reason for this may be that a strong background in statistics as well as in water chemistry is needed to successfully apply some of the more sophisticated statistical techniques during water-quality studies. Generally, for investigation of water chemistry, Hem (1985) suggested the use of statistics as a means of testing and verifying theories instead of simply creating theories from statistical data.

Statistical techniques are the most useful and appropriate when a large data base of observations is available. Simple averaging or determination of frequency distributions are widely used in water-analysis interpretation. Both are good techniques in establishing the background water-quality data often needed for identification of a salinization process. Together with an average value, a maximum background value is often given, which is then used to identify anomalously high concentrations caused by mixing with a salinization source. Frequency distributions are used on large data bases to identify any outliers or to determine the number of data populations that make up the total data set. Outliers may indicate contamination, faulty analyses, or data points not representative of the rest of the data. Analyses derived from wells that produce from different aquifers or from wells located in different geographic areas may result in more than one grouping of the data. This is important when sources of chemical changes are investigated, because these data groups may be unrelated to each other.

During mixing between a fresh water and a salt water, absolute concentrations vary to a high degree, whereas constituent ratios may vary relatively little. With only one potential salinization source present, absolute concentrations will most often provide enough information for identification of the salinization process, as concentrations exceed normal background levels. If more than one potential salinization source is present and a significant difference in constituent concentrations exists between them, absolute concentrations may still suffice to identify the actual source of mixing, as the concentration of a particular parameter may increase significantly and stand out. More often, however, absolute concentrations will not allow positive identification of the one true source of two or more potential sources because of overlapping concentrations, especially in the case of highly diluted mixing waters. In such cases, concentration ratios may have to be used for separation of endmembers. Na/Cl and Br/Cl are two ratios that are known to work well in a number of salinization scenarios (see also chapters 3.2 and 3.4), but which ratio works best will depend on the composition of the individual endmembers involved.

When examining scattergrams of major ionic constituents versus chloride for various potential endmembers of salinization (for example, sea water, halite-dissolution brines, and oil-field brines), it becomes apparent that some sources are more variable than others. Samples from halite solution (Fig. 22) or from sea-water intrusion (Fig. 35) show relatively little scatter, indicating little variation in end-member chemistry. When plotting oil-field brines together from different areas (Fig. 47), in contrast, much more scatter can be observed, reflecting the different origins and mechanisms of concentration of these brines. From this it is apparent that differentiation between sea water and oil-field brines or halite dissolution and oil-field brines cannot be done using a single, universal chemical constituent or constituent ratio. Instead, constituents to be used for differentiating between these brines will vary from one location to another in the same manner as the chemistry of the oil-field brine endmembers changes. One way to determine which constituents work best for distinguishing between salinization sources in any given case would be plotting or calculating ratios by trial and error until a good separation has been found. A more efficient way may be provided by Stepwise Discriminant Analysis (SDA), which is a statistical technique that identifies variables that distinguish between two or more predetermined groups of cases (Dixon and others, 1981). This technique was used successfully by Hitchon (1984) to group formation waters in the Western Canada

Sedimentary Basin, by Hawkins and Motyka (1984) to identify the origin of mineral springs in parts of the Copper River Basin, Alaska, and by Novak and Eckstein (1988) to differentiate salt water derived from road salt from brine samples derived from oil fields in Ohio. In terms of water chemistry, SDA can be used to determine those chemical constituents or constituent ratios that are most useful for distinguishing between given groups of waters, such as oil-field waters and sea water. When dealing with contamination of fresh water by salt water, the main mechanism changing the chemistry of the fresh water is mixing, which can also be considered as dilution of the salt water. In the case of dilution, absolute concentrations vary to a high degree, whereas concentration ratios change relatively little. Therefore, as long as the solution doesn't get too diluted and doesn't take on the ratio characteristics of the uncontaminated water, concentration ratios are generally better tracers of salt-water sources than absolute concentrations and were used in the following discussion of SDA.

In SDA, each predetermined group of samples consists of a number of chemical analyses made using a variety of parameters, for example, major ions. Any parameter or, in this case, parameter ratio (for example, Ca/Cl, Ca/SO_4, Mg/Cl, etc.) that is specified will be used during a SDA run to calculate mean values representative of each group. The difference between the groups of interest is then expressed in a linear function using the differences between the group means of each parameter. The ratio (for example, Ca/Cl) that best separates the groups, that is, for which mean values are most distinct between the two groups, is incorporated into this equation first. During this step a certain percentage of the individual analyses in each group will be assigned correctly to their preassigned groups, that is, parameter ratios are closer to the respective group mean than to the mean of the opposing group. However, in most cases, some of the analyses will be assigned to the opposite group, depending on the degree of difference between the given groups. In the next step, SDA will attempt to correctly assign the remaining analyses to their preassigned groups by incorporating another ratio in combination with the previously determined variable into the equation. The addition of ratios (variables) will continue in successive steps until a maximum number of analyses was assigned correctly to its preassigned group and no further improvement in sample assignment can be achieved by incorporating additional ratios into the equation. Depending on the similarity between the groups, any amount of ratios will be determined, that is, two groups that are relatively similar may necessitate inclusion of many ratios for separation of their group members into the assigned groups, whereas two very different groups may be distinguishable by just one or two ratios. The variables determined by SDA can then be used for further study.

To illustrate the usefulness of SDA for identification of ratios that distinguish between given groups, several data sets of water chemistry were compiled from the published literature. Data sets include oil-field brines, halite-solution brines, sea-water intrusion samples, and ground water. Chemical analyses were grouped according to brine type and location and then used in a variety of combinations to determine ratios that best separated these groups. Ratios used were those identified as useful in salinization studies (Table 12) and for which data were available (Ca, Mg, Na, K, HCO_3, SO_4, Cl, Br, and I): scenarios tested were (a) oil-field brine versus oil-field brine, (b) oil-field brine versus sea-water intrusion, (c) oil-field brine versus halite-solution brine, and (d) a known case of ground-water contamination by oil-field brine. The SDA software used in the following example is part of the BMDP software package, which is commercially available from the license holder to be run on mainframe computers or PC's.

SDA was performed on all 15 possible combinations between any two of six oil-field brine groups, representing brines from Texas, Louisiana, California, Oklahoma, Ohio, and Canada. Among the first three ratios selected during each run, Ca/Cl, Mg/Cl, Na/Cl, Br/Cl, and I/Cl were the ones most frequently identified (Table 15). It is probably reasonable to assume that combinations of these ratios will also provide good separation power between brines of the same type (oil-field or deep-basin brines) in other areas and that the other ratios may be useful only on a site-specific basis.

Solution of halite (chapter 3.2) and sea-water intrusion (chapter 3.3) each produce brine and saline ground water of relatively uniform chemical character (Figs. 22 and 35, respectively). Therefore, where local samples from sea-water intrusion or halite solution are not readily available, samples from other areas can be used as hypothetical endmembers with reasonable accuracy. This was done for identification of ratios that separate oil-field brine from sea-water intrusion brine in California, Texas, and Louisiana, and from halite-solution brine in Texas, Pennsylvania, and West Virginia. In the cases of sea-water intrusion versus oil-field brine, combinations of 7 (out of a possible 12) different ratios were determined as the best four ratios in the three test runs (Fig. 80), whereas in the case of halite solution versus oil-field brine, 10 ratios were identified (Fig. 81). Some ratios appear again as more useful than others, such as the Br/Cl ratio in the case of halite solution versus oil-field brine, or the HCO_3/Cl ratio in the case of sea-water intrusion versus oil-field brine. It should be emphasized here that SDA determines ratios from a strictly statistical point of view and not from a geochemical point of view. As pointed out in chapter 4, bicarbonate is not a conservative constituent in ground water, as its concentration is affected more by interactions between aquifer material and CO_2 with water than by mixing. Therefore, although the HCO_3/Cl ratio allows good separation between some oil-field brines and sea-water intrusion samples, the ratio may be of little help in identifying the source of salinity in water contaminated by any of these two sources. The same consideration should be given to any variable determined through statistical methods. In the case of Br/Cl ratios, geochemical considerations support the statistical evaluation, as discussed in chapters 3.4, and 4.

Plotting of step ratios 3 and 4 in bivariate plots, as done in figures 80 and 81, does not follow the SDA logic, as these ratios are determined after and in combination with ratios determined in step 1 and in step 2. Nevertheless, combining these ratios in bivariate plots can support group separations. Once the ratios are known that best separate potential endmembers of salinization, these ratios can possibly then be used to identify which endmember is the true source of salinity in a contaminated ground water. This can be done graphically or through the SDA feature of checking individual analyses for their similarity to endmember compositions (group means). First, SDA checks individual samples within each endmember group for their representativeness of that group, thus enabling the researcher to check the initial

Table 15. Listing of constituent ratios that separate best brines from Texas, Louisiana, Oklahoma, California, Ohio, and Canada, as determined through Stepwise Discriminant Analysis. Ratios were determined by individual runs of any combination between two brine groups, totaling 15 combinations between the six areas.

Ratio	Sequence Sequence of Selection [*]
Ca/Cl	1, 1, 1, 1, 1, 2, 2
Mg/Cl	1, 1, 1, 3, 3
Na/Cl	1, 1, 2
Br/Cl	1, 1, 2
I/Cl	2, 3, 3, 3, 3, 3
HCO_3/Cl	1, 2, 2
Ca/Mg	1, 2
K/Br	1, 3
Na/K	2, 3, 3
SO_4/(Na+K)	2, 2
Ca/Br	2, 3
(Br/Cl)/(Ca/Mg)	2, 3
SO_4/Cl	2
Na/Mg	2
Ca/K	3
Ca/SO_4	3
(Ca+Mg)/SO_4	3

Explanation:

1   Selected as the best ratio during any one run (step ratio #1), providing the single-most separation between two groups.
2   Selected as the second ratio after step ratio #1 during any run, providing improved separation between two groups in combination with step ratio #1.
3   Selected as the third ratio after step ratios #1 and #2 during any run, providing further improvement of separation in combination with step ratios #1 and #2.
*   Of the 15 combinations between the six data sets, Ca/Cl was selected five times as the first step ratio and two times as the second step ratio.

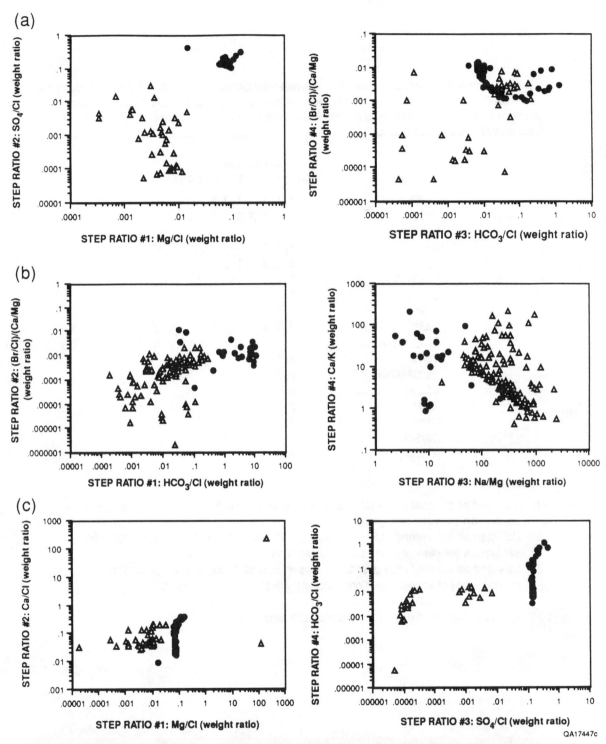

Figure 80. Bivariate plots of ratios determined by application of Stepwise Discriminant Analysis as the statistically best ratios to distinguish sea-water intrusion (solid dots) from oil-field brines (open triangles). Ratios change according to the composition of oil-field brines, derived from (a) California (data from Gullikson and others, 1961), (b) Texas (data from Kreitler and others, 1988), and (c) Louisiana (data from Dickey and others, 1972).

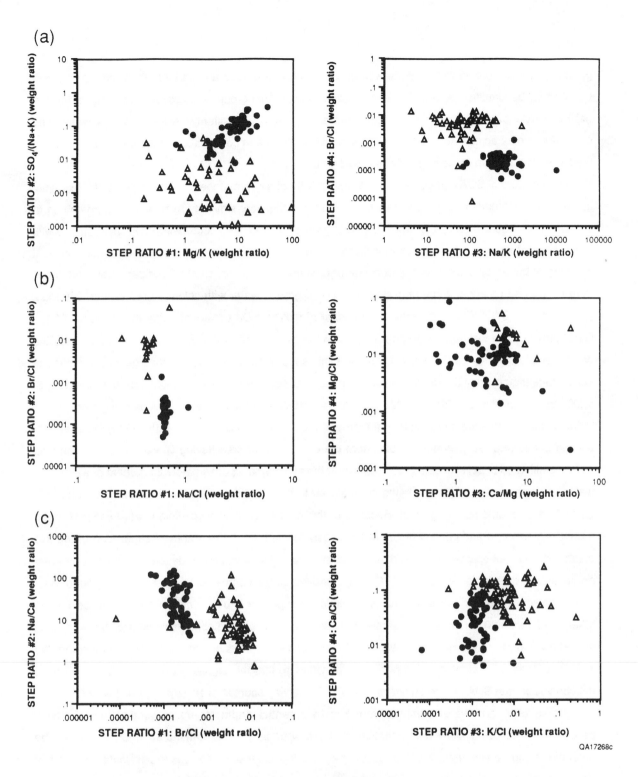

Figure 81. Bivariate plots of ratios determined by application of Stepwise Discriminant Analysis as the statistically best ratios to distinguish halite-solution brine (solid dots) from oil-field brines (open triangles). Ratios change according to the composition of oil-field brines, derived from (a) Texas (data from Kreitler and others, 1988), (b) Pennsylvania (data from Poth, 1962), and (c) West Virginia (data from Hoskins, 1947).

grouping. If initial grouping was done well and/or if the two groups are distinct, each individual group member will be identified by SDA as being representative of that group. If, however, an overlap of the two groups exists because of poor grouping and/or of chemical similarity, some samples within any endmember may actually turn out to be more similar to the mean composition of the other group than to the mean composition of the originally assigned group.

This feature of SDA can also be used to test individual group members of a third group (for example, apparently contaminated ground water) regarding their chemical similarity to potential endmembers of contamination (the two groups for which separating variables had been determined). This is illustrated with a test case of known oil-field pollution in Illinois, where salt-water disposal into pits and indiscriminant dumping of brine has caused local ground-water contamination (Lehr, 1969). Considering oil-field brines from Illinois as one endmember of mixing and Illinois' ground water with chloride greater than 250 mg/L (retrieved from STORET) as the other endmember of mixing, SDA determined that the ratios Na/Cl and Ca/Cl provide the most separation power between the two groups, with Ca, Mg, Na, SO_4, and Cl as the only available parameters. Figure 82a illustrates this separation, with oil-field brines generally plotting at lower ratios than most of the ground-water samples. Water samples from the contaminated area having Cl >200 mg/L (data from Van Biersel, 1985, and Stafford, 1987) were identified by SDA as being more similar to the oil-field brine endmember than to the ground-water endmember and plot within the general area of oil-field brines (Fig. 82a). Water samples from the contaminated area having Cl <200 mg/L, in contrast, were identified as being similar to ground water. At the same time, approximately 20 percent of all ground-water samples were classified as being more similar to the oil-field endmember than to the mean of the ground-water endmember (Fig. 82b). Because of this overlap of possibly oil-field related samples in the ground-water group and oil-field brines in the second endmember, the graphical separation of the two endmembers is not optimal and should be retried after careful examination of data within the ground-water group, that is, data retrieved from STORET and classified for the purpose of this test case as ground water should be scrutinized for possible inclusion of oil-field samples or of other contaminated samples. On the other hand, these 20 percent of the samples may contain some oil-field contamination and these samples should be reevaluated to determine whether they do show evidence of contamination. The fact that the known contaminated samples were positively identified so that this significant overlap existed supports the conclusion that SDA can be an effective tool for identifying sources of salinity in ground water.

Using SDA and water samples from known potential endmembers of salinization, chemical parameters that could allow differentiation of these sources in a contaminated fresh water can be determined prior to field investigations, possibly eliminating expensive analyses of parameters that are of little use.

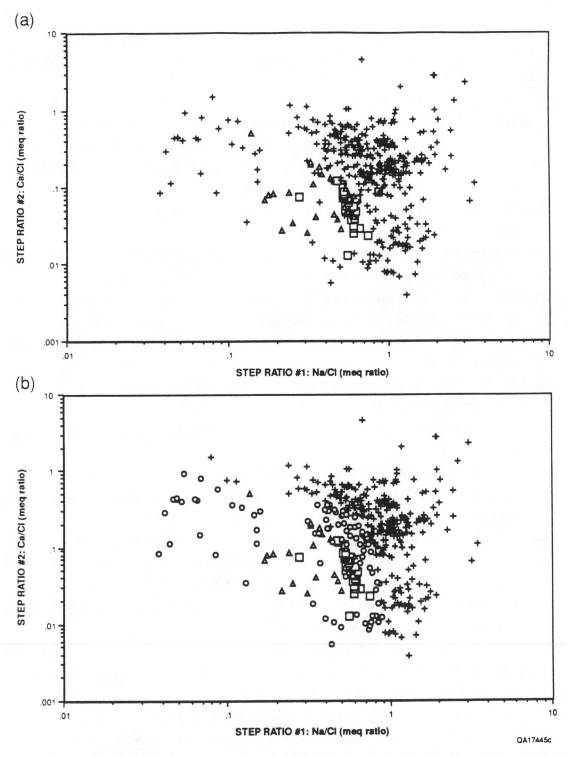

Figure 82. Bivariate plot of Na/Cl ratios versus Ca/Cl ratios for Illinois ground water (crosses; from STORET; Cl >250 mg/L), Illinois oil-field brines (open squares; data from Meents and others, 1952), and ground water contaminated by oil-field brine (open triangles; data from Stafford, 1987, and Van Biersel, 1985). Ratios determined through the use of Stepwise Discriminant Analysis (SDA) effectively separate ground water from brines (a). Statistical evaluation of individual analysis using SDA identified all contaminated water samples and 20 percent of the ground-water samples (open dots) as more similar to brines than to ground water (b).

229

7. REFERENCES

Albin, D. R., and Breummer, L. B., 1988, Minnesota ground-water quality, *in* Moody, D. W., Carr, Jerry, Chase, E. B., and Paulson, R. W., compilers, National Water Summary 1986—Hydrologic events and groundwater quality: U.S. Geological Survey Water-Supply Paper 2325, p. 313–320.

Alderton, D. H. M., and Sheppard, S. M. F., 1977. Chemistry and origin of thermal waters from southwest England: Institute Mining and Metallurgy, Transactions, v. (B)86, p. 191–194.

Alling, H. L., 1928, The geology and origin of the Silurian salt of New York State: New York State Museum Bulletin 275, 139 p.

Anderson, M. P., and Berkebile, C. A., 1976, Evidence of salt-water intrusion in southeastern Long Island: Ground Water, v. 14, no. 5, p. 315–319.

Arad, Arnon, Kafri, Uri, and Fleisher, Ezra, 1975, The Na'aman springs, northern Israel—Salination mechanism of an irregular freshwater-seawater interface: Journal of Hydrology, v. 25, p. 81–104.

Arkansas Department of Pollution Control and Ecology, 1984, Arkansas water quality inventory report 1984: Little Rock, 495 p.

Ashworth, J. B., 1990, Evaluation of ground-water resources in El Paso County, Texas: Texas Water Development Board Report 324, 25 p.

Atkinson, S. F., Miller, G. D., Curry, D. S., and Lee, S. B., 1986, Salt Water Intrusion: Lewis Publishers, 390 p.

Aulenbach, D. B., 1980, Litigation in the contamination of individual water supplies by storage of highway de-icing salt: Proceedings, 5th National Ground-Water Quality Symposium, Robert S. Kerr Environmental Research Laboratory.

Ayers, M. A., and Pustay, E. A., 1988, New Jersey ground-water quality, *in* Moody, D. W., Carr, Jerry, Chase, E. B., and Paulson, R. W., compilers, National Water Summary 1986—Hydrologic events and groundwater quality: U.S. Geological Survey Water-Supply Paper 2325, p. 369–376.

Back, William, 1966, Hydrochemical facies and ground-water flow patterns in northern part of the Atlantic Coastal Plain: U.S. Geological Survey Professional Paper 498-A, 42 p.

Back, William, Hanshaw, B. B., Pyle, T. E., Plummer, L. N., Weidie, A. E., 1979, Geochemical significance of groundwater discharge and carbonate solution to the formation of Caleta Xel Ha, Quintana Roo, Mexico: Water Resources Research, v. 15, no. 6, p. 1521–1535.

Bahls, L. L., and Miller, M. R., 1975, Ground-water seepage and its effects on saline soils: Bozeman, Montana State University Joint Water Resources Research Center Report No. 66, 39 p.

Bain, G. L., 1970, Salty ground water in West Virginia with a discussion of the Pocatalico River Basin above Sissonville: U.S. Geological Survey Circular 11, 31 p.

Bair, E. S., and Digel, R. K., 1990, Subsurface transport of inorganic and organic solutes from experimental road spreading of oil-field brine: Ground Water Monitoring Review, Summer 1990, p. 94–105.

Baker, F. G., and Brendecke, C. M., 1983, Seepage from oil-field brine disposal ponds in Utah: Ground Water, v. 21, no. 3, p. 317–324.

Ball Associates, Ltd, 1965, Surface and shallow oil-impregnated rocks and shallow oil fields in the U.S.: U.S. Department of the Interior, Bureau of Mines, Monograph 12, 375 p.

Balsters, R. G. and Anderson, C., 1979, Water quality effects associated with irrigation in Kansas: Kansas Water News, v. 22, nos. 1 and 2, Winter 1979, p. 14–22.

Banks, H. O., and Richter, R. C., 1953, Sea-water intrusion into ground-water basins bordering the California coast and inland bays: American Geophysical Union, Transactions, v. 34, no. 4, p. 575–582.

Banner, J. L., Wasserburg, G. J., Dobson, P. F., Carpenter, A. B., and Moore, C. H., 1989, Isotopic and trace element constraints on the origin and evolution of saline groundwaters from central Missouri: Geochimica et Cosmochimica Acta, v. 53, p. 383–398.

Barksdale, H. C., 1940, The contamination of ground-water by salt water near Parlin, New Jersey: American Geophysical Union, Transactions, v. 21, p. 471–474.

Bassett, R. L., and Bentley, M. E., 1983, Deep brine aquifers in the Palo Duro Basin—regional flow and geochemical constraints: The University of Texas at Austin, Bureau of Economic Geology Report of Investigations No. 130, 59 p.

Baughman, W. T., and McCarty, J. E., 1974, Water resources of Wayne County: Mississippi Geol. Econ. Topogr. Survey Bull., no. 117, p. 241–293.

Bednar, G. A., 1988, Mississippi ground-water quality, in Moody, D. W., Carr, Jerry, Chase, E. B., and Paulson, R. W., compilers, National Water Summary 1986—Hydrologic events and groundwater quality: U.S. Geological Survey Water-Supply Paper 2325, p. 321–328.

Behl, Elizabeth, Davis, S. N., and Goldwitz, Joshua, 1987, Cl/Br ratios as an environmental tracer of anthropogenically altered waters: Geological Society of America, Abstract with Programs, p. 585.

Bentley, H. W., Phillips, F. M., and Davis, S. N., 1986, Chlorine-36 in the terrestrial environment, in Fritz, P., and Fontes, J. C., eds., Handbook of environmental isotope geochemistry, v. 2, The terrestrial environment: Elsevier, p. 427–480.

Benz, L. C., Mickelson, R. H., Sandoval, F. M., and Carlson, C. W., 1961, Ground-water investigations in a saline area of the Red River Valley, North Dakota: Journal of Geophysical Research, v. 66, no. 8, p. 2435–2443.

Berg, W. A., Cail, C. R., Hungerford, D. M., Naney. J. W., and Sample, G. A., 1987, Saline seep on wheatland in northwest Oklahoma, in Fairchild, D. M., ed., Ground water quality and agricultural practices: Lewis Publishers, Chelsea, Michigan, p. 265–272.

Bingham, J. W., and Rolston, J. L., 1978, Road salt storage and road network in Connecticut: U.S. Geological Survey Miscellaneous Field Studies Map MF-981-A.

Bluntzer, R. L., 1981, Investigation of soil-salinity problem at four sites west of Estelline, Hall County, Texas: Texas Department of Water Resources, Planning and Development Division Open-File Report 81-2, 23 p.

Bluntzer, R. L., 1982, Investigation of ground-water salinity and soil-salinity problems in the Brownfield Lakes—Red Onion Flats depression of Terry and Lynn Counties, Texas: Texas Department of Water Resources, Planning and Development Division Open-File Report 82-1, 96 p.

Boggess, D. H., 1970, The magnitude and the extent of salt-water contamination in the Caloosahatchee River between La Belle and Olga, Florida: U.S. Geological Survey, Florida Bureau of Geology Information Circular No. 62, part 2, p. 17–39.

Boggess, D. H., 1973, The effects of plugging a deep artesian well on the concentration of chloride in water in the water-table aquifer at Highland Estates, Lee County, Florida: U.S. Geological Survey Open-File Report No. 73003.

Bolke, E. L., and Price, Don, 1969, Hydrologic Reconnaissance of Curlew Valley, Utah and Idaho: Utah Department of Natural Resources Technical Publication No. 25, 40 p.

Boyd, F. M., and Kreitler, C. W., 1986, Hydrogeology of a gypsum playa, northern salt basin, Texas: The University of Texas at Austin, Bureau of Economic Geology, Report of Investigations No. 158, 37 p.

Breen, K. J., Angelo, C. G., Masters, R. W., and Sedam, A. C., 1985, Chemical and isotopic characteristics of brines from three oil- and gas-producing sandstones in eastern Ohio, with applications to the geochemical tracing of brine sources: U.S. Geological Survey Water-Resources Investigations Report 84-4314, 58 p.

Brennan, Robert, 1956, Sea-water intrusion in California—Preliminary chemical-quality study in the Manhattan Beach area, California: California Department of Water Resources, Bulletin No. 63, Appendix E, 34 p.

Brown, H. T., 1984, A case study of a salinity control program, Uinta Basin, Utah, *in* French, R. H., ed., Salinity in watercourses and reservoirs, Proceedings of the 1983 International Symposium on State-of-the-Art Control of Salinity, Salt Lake City, Utah: Butterworth Publishers, Boston, p. 275–284.

Brown, E. B., Skougstad, M. W., and Fishman, M. J., 1970, Methods for collection and analysis of water samples for dissolved minerals and gases: Techniques of water-resources investigations of the U.S. Geological Survey, Book 5, Chapter A1, 130 p.

Bruington, A. E., 1972, Salt-water intrusion into aquifers: Water Resources Bulletin, v. 8, no. 1, p. 150–160.

Bubeck, R. C., Diment, W. H., Deck, B. L., Baldwin, A. L., and Lipton, S. D., 1971, Runoff of deicing salt—effect on Irondequoit Bay, Rochester, New York: Science, v. 175, p. 1128–1132.

Buetner, E. C., and Charles, E. G., 1982, Large volume loss during cleavage formation, Hamburg sequence, Pennsylvania: Geol. 13, p. 803–805

Burnitt, S. C., 1963, Reconnaissance of soil damage and ground-water quality, Fisher County, Texas: Texas Water Commission, Memorandum Report No. 63-02, 45 p.

Burnitt, S. C., Adams, J. B., Rucker, O. C., and Porterfield, H. C., 1963, Effects of surface-disposal of oil-field brine on the quality and development of ground water in the Ogallala Formation, High Plains of Texas, and a hydrologic study of brine disposal into surface pits, High Plains of Texas: Texas Water Commission, 115 p.

Cagle, J. W., 1969, Availability of groundwater in Wayne County, Iowa: Iowa Geological Survey Water Atlas No. 3, 33 p.

Canter, L. W., 1987, Nitrates and pesticides in ground water: an analysis of a computer-based literature search, *in* Fairchild, D. M., ed., Ground water quality and agricultural practices, Lewis Publishers, Chelsea, Michigan, p. 153–174.

Carlston, C. W., and Graeff, G. D., Jr., 1955, Ground-water resources of the Ohio River Valley in West Virginia: West Virginia Geological and Economic Survey, v. XXII, part III.

Carothers, W. W., and Kharaka, Y. K., 1978, Aliphatic acid anions in oil-field waters—implications for origin of natural gas: American Association of Petroleum Geologists Bulletin, v. 62, no. 12, p. 2441–2453.

Carpenter, A. B., and Darr, J. M., 1978, Relationship between groundwater resources and energy production in southwestern Missouri: Rolla, Missouri Water Resources Research Center, Completion Report, 113 p.

Chandler, R. V., Moore, J. D., and Gillett, B., 1985, Ground-Water Chemistry and salt-water encroachment, Southern Baldwin County, Alabama: Geological Survey of Alabama, Water Resources Division, Bulletin 126.

Chaudhuri, S., 1978, Strontium isotopic composition of several oil-field brines from Kansas and Colorado: Geochimica et Cosmochimica Acta, v. 42, p. 329–331.

Childs, K. E., 1970, History of the salt, brine and paper industries and their probable effect on the ground water quality in the Manistee Lake area of Michigan: Hydrological Survey Division, Water Resources Commission, Bureau of Water Management, Michigan Department of Natural Resources, 69 p.

Clarke, J. S., and McConnell, J. B., 1988, Georgia ground-water quality, *in* Moody, D. W., Carr, Jerry, Chase, E. B., and Paulson, R. W., compilers, National Water Summary 1986—Hydrologic events and groundwater quality: U.S. Geological Survey Water-Supply Paper 2325, p. 215–222.

Clayton, R. N., Friedman, I., Graf, D. L., Mayeda, T. K., Meents, W. F., and Shimp, N. F.,1966, The origin of saline formation waters: I. Isotopic composition, Journal of Geophysical Research 71, p. 869–3889

Collins, A. G., 1967, Geochemistry of some Tertiary and Cretaceous age oil-bearing formation waters: Environmental Science and Technology, v. 1, no. 9, p. 725–730.

Collins, A. G., 1969, Chemistry of some Anadarko Basin brines containing high concentrations of iodide: Chemical Geology, v. 4, no. 1–2, p.169–187.

Collins, A. G., 1974, Saline ground waters produced with oil and gas: U.S. Environmental Protection Agency, Office of Research and Development, Technology Series EPA-660/2-74-010, 68 p.

Cooper, H. H., Jr., Kohout, F. A., Henry, H. R., and Glorer, R. E., 1964, Sea water in coastal aquifers, relation of salt water to fresh water: U.S. Geological Survey, Water-Supply Paper 1613C, 84 p.

Core Laboratories, Inc., 1972, Chemical analyses of saline water: Texas Water Development Board, Report 157, v. 2, 378 p.

Cotecchia, V., Tazioli, G. S., Magri, G. 1974, Isotropic measurements in research on seawater ingression in the carbonate aquifer of the Salentine Peninsula, Southern Italy, Isotope Techniques in Groundwater Hydrology 1974: International Atomic Energy Agency, Vienna. p. 445–463.

Cotton, J. E., and Butterfield, D., 1988, Vermont ground-water quality, in Moody, D. W., Carr, Jerry, Chase, E. B., and Paulson, R. W., compilers, National Water Summary 1986—Hydrologic events and groundwater quality: U.S. Geological Survey Water-Supply Paper 2325, p. 501–508.

Craig, Harmon, 1961, Isotopic variations in meteoric waters: Science, v. 133, no. 3465, p. 1702–1703.

Craig, J. R., 1970, Saline Waters—Genesis and relationship to sediments and host rocks, in Mattox, R. B., ed., Ground-water salinity: Symposium, 46th Annual Meeting of the Southwestern and Rocky Mountain Division of the American Association of the Advancement of Science, April 23–24, Las Vegas, New Mexico, p. 3–29.

Crain, L. J., 1966, Ground-water resources of the Jamestown Area, New York: State of New York Conservation Department, Water Resources Commission Bulletin No. 58.

Crain, L. J., 1969, Ground-water pollution from natural gas and oil production in New York: U.S. Geological Survey Report of Investigation No. 5, 15 p.

Custer, S. G., 1979, Regional identification of saline-seep recharge areas, Stillwater and Yellowstone Counties, Montana: Montana Water Resources Research Center Report No. 102, 58 p.

Custodio, E., 1987, Hydrogeochemistry and tracers, in Custodio, E., Ground-water problems in coastal areas: Studies and reports in hydrology no. 45, UNESCO, p. 213–269.

Davis M. E., and Gordon, J. D., 1970, Records of water levels and chemical analysis from selected wells in parts of the Trans-Pecos region, Texas, 1965–68: Texas Water Development Board Report 114, 49 p.

Denver, J. M., 1988, Delaware ground-water quality, in Moody, D. W., Carr, Jerry, Chase, E. B., and Paulson, R. W., compilers, National Water Summary 1986—Hydrologic events and groundwater quality: U.S. Geological Survey Water-Supply Paper 2325, p. 199–204.

DeSitter, L. U., 1947, Diagenesis of oil-field brines: American Association of Petroleum Geologists Bulletin, v. 31, p. 2030–2040.

Deverel, S. J., and Gallanthine, S. K., 1989, Relation of salinity and selenium in shallow ground water to hydrologic and geochemical processes, western San Joaquin Valley, California: Journal of Hydrology, v. 109, p. 125–149.

Dickey, P. A., Collins, A. G., and Fajardo, Ivan, 1972, Chemical composition of deep formation waters in Southwestern Louisiana; American Association of Petroleum Geologists Bulletin, v. 56, no. 8, p. 1530–1533.

Diment, W. H., Bubeck, R. C., and Deck, B. L., 1973, Effects of deicing salt on the waters of the Irondequoit Bay drainage basin, Monroe County, New York, in Coogan, A. H., ed., Fourth Symposium on Salt: Northern Ohio Geological Society, v. 2., p. 391–405.

Dion, N. P., and Sumioka, S. S., 1984, Sea-water intrusion into coastal aquifer in Washington, 1978: U.S. Geological Survey, Water Resources Division, Water-Supply Bulletin, v. 56, 13 p.

Dixon, W. J., Brown, M. B., Engelman, L., Frne, J. W., Hill, M. A., Jennrich, R. I., and Toporek, J. D., 1981, BMDP statistical software: Berkeley, University of California Press, 725 p.

Doneen, L. D., 1966, Factors contributing to the quality of agricultural waste waters in California, in Doneen, L. D., ed., Symposium on Agricultural Waste Waters: University of California, Davis, Water Resources Center Report 10, p. 6–9.

Donovan, J. J., Sonderegger, J. L., and Miller, M. R., 1981, Investigations of soluble salt loads, controlling mineralogy, and factors affecting the rates and amounts of leached salts: Montana Bureau of Mines, Water Resources Research Center Report No. 120, 61 p.

Drever, J. I., 1982, The geochemistry of natural waters: Prentice-Hall, Inc., Englewood Cliffs, New Jersey, 388 p.

Duce, R. A., Winchester, J. W., and Van Nahl, T. W., 1965, Iodine, bromine, and chlorine in the Hawaiian marine atmosphere: Journal of Geophysical Research, v. 70, no. 8, p. 1775–1799.

Dunrud, C. R., and Nevins, B. B., 1981, Solution mining and subsidence in evaporite rocks in the United States: U.S. Geological Survey Miscellaneous Investigations Series Map 1298.

Dutton, A. R., Fisher, B. C., Richter, B. C., and Smith, D. A., 1985, Hydrologic testing in the salt-dissolution zone of the Palo Duro Basin, Texas Panhandle: The University of Texas at Austin, Bureau of Economic Geology Open-File Report OF-WTWI-1985-3, 15 p.

Dutton, A. R., Richter, B. C., and Kreitler, C. W., 1989, Brine discharge and salinization, Concho River watershed, West Texas: Ground Water, v. 27, no. 3, p 375–383.

Dutton, A. R., and Richter, B. C., 1990, Regional geohydrology of the Gulf Coast Aquifer in Matagorda and Wharton Counties, Texas: Development of a numerical model to estimate the impact of water-management strategies: The University of Texas at Austin, Bureau of Economic Geology, Report prepared for Lower Colorado River Authority under Contract Number IAC (88-89)0910, 116 p.

Edelmann, Patrick, and Buckles, D. R. 1984, Quality of ground water in agricultural areas of the San Luis Valley, south central Colorado: U.S Geological Survey Water-Resources Investigations Report 83-4281, 37 p.

Edmunds, W. M., Andrews, J. N., Burgess, W. G., Kay, R. L. F., and Lee, D. J., 1984, The evolution of saline and thermal groundwaters in the Carnmenellis granite: Mineralogical Magazine, v. 48, p. 407–424.

Edmunds, W. M., Kay, R. L. F., and McCartney, R. A., 1985, Origin of saline groundwaters in the Carnmenellis granite—natural processes and reaction during Hot Dry Rock reservoir circulation: Chemical Geology, v. 49, p. 287–301.

Edmunds, W. M., Kay, R. L. F., Miles, D. L., and Cook, J. M., 1987, The origin of saline ground waters in the Carnmenellis Granite, Cornwall (U.K.)—further evidence from minor and trace elements, in Fritz, P., and Frape, S. K., eds., Saline water and gases in crystalline rocks: Geological Association of Canada Special Paper 33, p. 127–143.

Effertz, R. J., Sidebottom, W. K., and Turley, M. D., 1984, Measures for reducing return flows from the Wellton-Mohawk irrigation and drainage district, in French, R. H., ed., Salinity in watercourses and reservoirs, Proceedings of the 1983 International Symposium on State-of-the-Art Control of Salinity, Salt Lake City, Utah: Butterworth Publishers, Boston, p. 305–314.

Engberg, R. A., and Druliner A. D., 1988, Nebraska ground-water quality, in Moody, D. W., Carr, Jerry, Chase, E. B., and Paulson, R. W., compilers, National Water Summary 1986—Hydrologic events and groundwater quality: U.S. Geological Survey Water-Supply Paper 2325, p. 347–354.

Exner, M. E. and Spalding, R. F., 1979, Evaluation of contaminated groundwater in Holt County, Nebraska: Water Resources Research, v. 15, no. 1, p. 137–147.

Fairchild, H. L., 1935, Genesee Valley hydrography and drainage: Proceedings of the Rochester Academy of Science, v. 7, p. 157–188.

Farrar, C. D., and Bertoldi, G. L., 1988, Region 4, Central Valley and Pacific Coast Ranges, in Back, William, Rosendheim, J. S., and Seaber, P. R., eds., Hydrogeology: Geological Society of America, The Geology of North America, v. O-2, p. 59–67.

Feast, C. F., 1984, Meeker dome salinity investigation, in French R. H., ed., Salinity in watercourses and reservoirs: Butterworth Publishers, Boston, p. 359–365.

Feltz, H. R., and Hanshaw, B. B., 1963, Preparation of water sample for carbon-14 dating: U.S. Geological Survey, Geological Survey Circular 480, p. 1–3.

Ferreira, R. F., Cannon, M. R., Davis, R. E., Shewman, F., and Arrigo, J. L., 1988, Montana ground-water quality, in Moody, D. W., Carr, Jerry, Chase, E. B., and Paulson, R. W., compilers, National Water Summary 1986—Hydrologic events and groundwater quality: U.S. Geological Survey Water-Supply Paper 2325, p. 165–172.

Feth, J. H., and others, 1965, Preliminary map of the conterminous United States showing depth to and quality of shallowest ground water containing more than 1,000 parts per million dissolved solids: U.S. Geological Survey Hydrologic Atlas HA-199, 2 sheets.

Field, Richard, Struzeski, E. J., Jr., Masters, H. E., and Tafuri, A. N., 1973, Water pollution and associated effects from street salting: U.S. Environmental Protection Agency, National Environmental Research Center Report EPA-R2-73-257, 48 p.

Fink, B. E., 1965, Investigation of Ground-and Surface-Water Contamination Near Harrold, Willbarger County, Texas, Texas Water Commission Report LD-0365, 23 p.

Fireman, M., and Hayward, H. E., 1955, Irrigation water and saline and alkali soils, in U.S. Department of Agriculture, Water, The Yearbook of Agriculture, 1955, p. 321.

Fogg, G. E., 1980, Salinity of Formation Waters, in Kreitler, C. W., and others, Geology and Geohydrology of the East Texas Basin: The University of Texas at Austin, Bureau of Economic Geology Geological Circular 80-12, p. 68–72.

Fogg, G. E., Kreitler, C. W., and Dutton, S. P., 1980, Hydrologic Stability of Oakwood Dome, in Kreitler, C. W., and others, Geology and Geohydrology of the East Texas Basin: The University of Texas at Austin, Bureau of Economic Geology Geological Circular 80-12, p. 30–31.

Fogg, G. E., and Kreitler, C. W., 1980, Impacts of Salt-Brining on Palestine Salt Dome—Geology and Geohydrolgy of the East Texas Basin: The University of Texas at Austin, Bureau of Economic Geology Geological Circular 80-12, p. 46–54.

Fogg, G. E., and Kreitler, C. W., 1982, Ground-water hydraulics and hydrochemical facies in Eocene aquifers of the East Texas Basin: The University of Texas at Austin, Bureau of Economic Geology Report of Investigations 127, 75 p.

Food and Agriculture Organization of the United Nations, 1973, Irrigation, drainage and salinity: Hutchinson & Co (publishers) LTD, London, 510 p.

Framji, K. K., 1976, Irrigation and salinity—a world-wide survey: International Commission on irrigation and drainage, New Delhi, India, 106 p.

Frape, S. K., and Fritz, P., 1982, The chemistry and isotopic composition of saline groundwaters from the Sudbury Basin, Ontario: Canadian Journal of Earth Sciences, v. 19, p. 645–661.

Frape, S. K., and Fritz, P., 1987, Geochemical trends for ground waters from the Canadian Shield, in Fritz, P., and Frape, S. K., eds., Saline water and gases in crystalline rocks: Geological Association of Canada Special Paper 33, p. 19–38.

Frape, S. K., Fritz, P., and McNutt, R. H., 1984, The role of water–rock interaction in the chemical evolution of groundwaters from the Canadian Shield: Geochimica et Cosmochimica Acta, v. 48, p. 1617–1627.

Freeze, R. A., and Cherry, J. A., 1979, Groundwater: Prentice-Hall, Inc., Englewood Cliffs, New Jersey, 604 p.

Fritz, P., and Frape, S. K., 1982, Saline groundwaters in the Canadian Shield—a first overview: Chemical Geology, v. 36, p. 179–190.

Fryberger, J. S., 1972, Rehabilitation of a brine-polluted aquifer: U.S. Environmental Protection Agency, Office of Research and Monitoring, Technology Series Report EPA-R2-72-014, 61 p.

Fuhriman, D. K., and Barton, J. R., 1971, Ground-water pollution in Arizona, California, Nevada, and Utah: U.S. Environmental Protection Agency, Office of Research and Monitoring, Water Pollution Control Research Series Report No. 16060 ERU 12/71, 249 p.

Garner, Stuart, 1988, Making the most of field-measurable ground-water quality parameters: Ground Water Monitoring Review, v. 8, no. 3, p. 60–66.

Garrels, R. M., and MacKenzie, F. T., 1967, Origin of the chemical compositions of some springs and lakes, in Gould, R. F., ed., Equilibrium concepts in natural water systems: American Chemical Society, Advances in Chemistry Series 67, p. 222–242.

Garrels, R. M., and MacKenzie, F. T., 1971, Evolution of Sedimentary Rocks: Norton, New York, p. 272

Garza, Sergio, and Wesselman, J. B., 1959, Geology and ground-water resources of Winkler County, Texas: Texas Board of Water Engineers Bulletin 5916, 200 p.

Gascoyne, M., Davison, C. C., Ross, J. D., and Pearson, R., 1987, Saline ground waters and brines in plutons in the Canadian Shield, in Fritz, P., and Frape, S. K., eds., Saline water and gases in crystalline rocks: Geological Association of Canada Special Paper 33, p. 53–68.

Gass, T. E., Lehr, J. H., and Heiss, H. W., Jr., 1977, Impact of abandoned wells on ground water: U.S. Environmental Protection Agency, Office of Research and Development, Robert S. Kerr Environmental Research Laboratory, Ecological Research Series EPA-600/3-77-095, 52 p.

Geraghty, Miller, van der Leeden, and Troise, 1973, Water Atlas of the United States: Water Information Center, Port Washington, New York.

Gillespie, J. B., and Hargadine, G. D., 1981, Saline ground-water discharge to the Smoky Hill River between Salina and Abilene, Central Kansas: U.S. Geological Survey Water-Resources Investigations No. 81-43, 72 p.

Gogel, Tony, 1981, Discharge of salt water from Permian rocks to major stream-aquifer systems in Central Kansas: Kansas Geological Survey, The University of Kansas at Lawrence, Chemical Quality Series 9, 60 p.

Goldberg, E. D., Broecker, W. S., Gross, M. G., and Turekian, K. K., 1971, Marine chemistry, in Radioactivity in the marine environment: Washington, D.C., National Academy of Sciences, p. 137–146.D

Grigsby, C. O., Tester, J. W., Trujillo, P. E., Counce, D. A., Abbott, J., Holley, C. E., and Blatz, L. A., 1983, Rock–water interactions in hot dry rock geothermal systems: field investigations of in situ geochemical behavior: Journal of Volcanology and Geothermal Research, v. 15, p. 101–136.

Gullikson, D. M., Caraway, W. H., and Gates, G. L., 1961, Chemical analysis and electrical resistivity of selected California oil-field waters: U.S. Department of the Interior, Bureau of Mines Report of Investigations No. 5736, 21 p.

Gustavson, T. C., 1979, Salt Dissolution, in Dutton, S. P., and others, eds., Geology and geohydrology of the Palo Duro Basin, Texas Panhandle, A report on the progress of nuclear waste isolation feasibility studies (1978): The University of Texas at Austin, Bureau of Economic Geology, Geological Circular 79-1, p. 87–99.

Gustavson, T. C., Finley, R. J., and McGillis, K. A., 1980, Regional dissolution of Permian salt in the Anadarko, Dalhart, and Palo Duro Basins of the Texas Panhandle: The University of Texas at Austin, Bureau of Economic Geology Report of Investigations No. 106, 40 p.

Gustavson, T. C., Holliday, V. T., and Schultz, G. E., 1990, Introduction, in Gustavson, T. C., ed., Tertiary and Quaternary stratigraphy and vertebrate paleontology of parts of northwestern Texas and eastern New Mexico: The University of Texas at Austin, Bureau of Economic Geology Guidebook 24, p. 1–21.

Hale, Allan, 1973, Salt water vapor discharges and the environment, in Coogan, A. H., ed., Fourth Symposium on Salt: Northern Ohio Geological Society, v. 2., p. 387–389.

Hamilton, P. A., and Denver, J. M., 1990, Effects of land use and ground-water flow on shallow ground-water quality, Delmarva Peninsula, Delaware, Maryland, and Virginia: Ground Water, v. 28, no. 5, p. 789.

Hamlin, H. S., Smith, D. A., and Akhter, M. S., 1988, Hydrogeology of Barbers Hill salt dome, Texas Coastal Plain: The University of Texas at Austin, Bureau of Economic Geology Report of Investigations No. 176, 41 p.

Hanor, J. S., 1983, Fifty years of development of thought on the origin and evolution of subsurface sedimentary brines, in Boardman, S. J., ed., Revolution in the earth sciences—Advances in the past half-century: Dubuque, Kendall/Hunt, p. 99–111.

Hanor, J. S., and Workman, A. L., 1986, Distribution of dissolved volatile fatty acids in some Louisiana oil field brines, in F. E. Ingerson Festschrift; Part 1, Applied Geochemistry, v. 1, no. 1, p. 37–46.

Hanshaw, B. B., Back, William, and Deike, R. G., 1971, A geochemical hypothesis for dolomitization by ground water: Economic Geology, v. 66, p. 710–724.

Hanshaw, B. B., Back, William, Rubin, M., and Wait, R., 1965, Relation of Carbon-14 Concentrations to Saline Water Contamination of Coastal Aquifers: Water Resources Research, v. 1, p. 109–114.

Haque, S. M., 1989, Environmental effects of surface brine disposal in Golden Meadow Oil and Gas field area, Lafourche Parish, Louisiana [abs]: Geological Society of America, 1989 Annual Meeting, Abstracts with Programs, p. A187.

Hardie, L. A., and Eugster, H. P., 1970, The evolution of closed basin brines: Mineralogical Society of America Special Publication, No. 3, p. 273–290.

239

Hargadine, G. D., Balsters, R. G., and Luehring, Joann, 1979, Mineral intrusion in Kansas surface waters—a technical report: Kansas Water Office, Topeka, Kansas, 211 p.

Harrel, R. C., 1975, Water quality and salt-water intrusion in the Lower Neches River: Texas Journal of Science, v. 26, nos. 1–2, p. 107–117.

Havens, J. S., and Wilkins, D. W., 1979, Experimental salinity alleviation at Malaga Bend of the Pecos River, Eddy County, New Mexico: U.S. Geological Survey Water-Resources Investigations 80-4, 65 p.

Hawkins, D. B., and Motyka, R. J., 1984, A multivariate statistical analysis and chemical mass balance analysis of waters of the Copper River Basin, Alaska, in Hitchon, Brian, and Wallick, E. I., eds., Proceedings, First Canadian/American Conference on Hydrogeology—Practical applications of ground-water geochemistry: National Water Well Association, Worthington, Ohio, p. 238–245.

Hearne, G. A., Lindner-Lunsford, J., Cain, D., Watts, K. R., Robson, S. G., Tobin, R. L., Teller, R. W., Schneider, P. A., and Gearhart, M. J., 1988, Colorado ground-water quality, in Moody, D. W., Carr, Jerry, Chase, E. B., and Paulson, R. W., compilers, National Water Summary 1986—Hydrologic events and groundwater quality: U.S. Geological Survey Water-Supply Paper 2325, p. 181–190.

Helgeson, H. C., 1972, Chemical interaction of feldspars and aqueous solutions: in Feldspars, MacKenzie, W. S., and Zussman, J., Procedings, NATO Advanced Study Institute, Manchester University Press, p. 187–217.

Helgeson, J. O., 1990, Effects of agricultural and petroleum-production land use on water quality in the High Plains Aquifer, south-central Kansas: Ground Water, v. 28, no. 5, p. 795.

Hem, J. D., 1985, Study and interpretation of the chemical characteristics of natural water: U.S. Geological Survey Water-Supply Paper 2254, 263 p.

Hingston, F. J. and Gailitis, V., 1976: The Geographic Variation of Salt Precipitated over Western Australia: Australian Journal of Soil Resources, v. 14, p. 319–335.

Hiss, W. L., 1970, Acquisition and machine processing of saline water data from Southeastern New Mexico and Western Texas: U.S. Geological Survey, Water Resources Research, v. 6, no. 5, p. 1471–1477.

Hitchon, Brian, Levinson, A. A., and Reeder, S. W., 1969, Regional variations of river water composition resulting from halite solution, Mackenzie River Drainage Basin, Canada: Water Resources Research, v. 5, no. 6, p. 1395–1403.

Hitchon, Brian, 1984, Graphical and statistical treatment of standard formation water analyses, in Hitchon, Brian, and Wallick, E. I., eds., Proceedings, First Canadian/American Conference on Hydrogeology—Practical applications of ground-water geochemistry: National Water Well Association, Worthington, Ohio, p. 238–245.

Holland, T. W., and Ludwig, A. H., 1981, Use of water in Arkansas, 1980: Arkansas Geological Commission Water Resources Summary No. 14, 30 p.

Hopkins, H. T., 1963, The effect of oil-field brines on the potable ground water in the Upper Big Pitman Creek Basin, Kentucky: Kentucky Geological Survey, Series X, Report of Investigations No. 4, 36 p.

Hoskins, H. A., 1947, Analyses of West Virginia brines: West Virginia Geological and Economic Survey, Report of Investigations No. 1, 22 p.

Howard, K. W. F., and Lloyd, J. W., 1983, Major ion characterization of coastal saline ground waters: Ground Water, v. 21, no. 4, p. 429–437.

Hubbert, M. K., 1940, The theory of ground-water motion, Journal of Geology, v. 48, p. 785–944.

Hunt, C. B., 1960, The Death Valley salt pan, a study of evaporites: U.S. Geological Survey Professional Paper 400-B, p. B456–B457.

Hunt, C. B., Robinson, T. W., Bowles, W. A., and Washburn, A. L., 1966, Hydrologic basin, Death Valley, California: U.S. Geological Survey Professional Paper 494-B, 138 p.

Hutchinson, F. E., 1973, Effect of deicing salts applied to highways on the contiguous environment, in Coogan, A. H., ed., Fourth Symposium on Salt: Northern Ohio Geological Society, v. 1., p. 427–434.

Ikeda, Kiyoji, 1967, Study on salt water intrusion into ground water—part 2. Geochemical research of salt in groundwater: Geological Survey of Japan, v. 18, no. 6, p. 15(393)–33(411).

International Atomic Energy Agency, 1981, Sampling of water for ^{14}C analysis: Isotope Hydrology Laboratory, Vienna.

Irwin, G. A., Bonds, J. L., 1988, Florida ground-water quality, in Moody, D. W., Carr, Jerry, Chase, E. B., and Paulson, R. W., compilers, National Water Summary 1986—Hydrologic events and groundwater quality: U.S. Geological Survey Water-Supply Paper 2325, p. 205–214.

Jacks, G., 1973, Chemistry of some ground waters in igneous rocks: Nordic Hydrology, v. 4, no. 4, p. 207–236.

Jensen, E. G., 1978, The Paradox Valley unit—The problem and the proposed solution: EOS, American Geophysical Union Transactions, v. 59, no. 12, p. 1064.

Johnson, B. A., and Riley, J. A., 1984, Price/San Rafael River Basin salinity investigation, in French, R. H., ed., Salinity in watercourses and reservoirs, Proceedings of the 1983 International Symposium on State-of-the-Art Control of Salinity, Salt Lake City, Utah: Butterworth Publishers, Boston, p. 407–416.

Johnson, K. S., Brokaw, A. L., Gilbert, J. F., Saberian, A., Snow, R. H., and Walters, R. F., 1977, Summary report on salt dissolution review meeting, March 29–30, 1977: Union Carbide Corporation, Nuclear Division, Office of Waste Isolation, Report Y/OWI/TM-31, 10 p.

Johnston, R. H., and Miller, J. A., 1988, Region 24, Southeastern United States, in Back, William, Rosendheim, J. S., and Seaber, P. R., eds., Hydrogeology: Geological Society of America, The Geology of North America, v. O-2, p. 229–236.

Jones, B. F., 1965, Geochemical evolution of closed basin water in the Western Great Basin, *in* Rau, J. L., ed., Second symposium on Salt: Northern Ohio Geological Society, v. 1, p. 181–200.

Jones, B. F., Vandenburgh, A. S., Truesdell, A. H., and Rettig, S. L., 1969, Interstitial brines in playa sediments: Chemical Geology, v. 4., p. 253–262.

Jones, P. H., and Hutchon, H., 1983, Road salt in the environment, *in* Schreiber, B. C., and Harner, H. L., eds., Sixth international symposium on salt, v. 2: Salt Institute, Alexandria, Virginia, p. 615–619.

Jonez, A. R., 1984, Controlling salinity in the Colorado River Basin, the arid West, *in* French, R. H., ed., Salinity in watercourses and reservoirs, Proceedings of the 1983 International Symposium on State-of-the-Art Control of Salinity, Salt Lake City, Utah: Butterworth Publishers, Boston, p. 337–347.

Jorgensen, D. G., 1968, An aquifer test used to investigate a quality of water anomaly: Ground Water, v. 6, no. 6, p. 18–20.

Jorgensen, D. G., 1977, Salt-water encroachment in aquifers near the Houston Ship Channel, Texas: U.S. Geological Survey Open-File Report 76-781, 45 p.

Jorgensen, D. G., 1981, Geohydrologic models of the Houston District, Texas: Ground Water, v. 19, no. 4, p. 418–428.

Kafri, U., and Arad, A., 1979, Current subsurface intrusion of Mediterranean seawater—a possible source of groundwater salinity in the rift-valley system, Israel: Journal of Hydrology, v. 44, p. 267–287.

Kaiser, W. R., Swartz, T. E., and Hawkins, G. J., 1991, Hydrology of the Fruitland Formation, San Juan Basin, *in* Ayers, W. B., and others, eds., Geologic and hydrologic controls on the occurrence and producibility of coalbed methane, Fruitland Formation, San Juan Basin: The University of Texas at Austin, Bureau of Economic Geology, topical report prepared for the Gas Research Institute under Contract No. 5087-214-1544, (GRI-91/0072), p. 195–241.

Kalka, M. S., Eckstein, Yoram, and Dahl, P. S., 1989, Investigation of the fate of heavy metals of oil field brines used as an anti-dust agent on roads in northeastern Ohio [abs.]: Geological Society of America, Abstracts with Programs, p. A243.

Kalkhoff, S. J., 1986, Brine contamination of fresh-water aquifers and streams in petroleum producing areas in Mississippi: U.S. Geological Survey Water-Resources Investigations Report 85-4117, 106 p.

Kantrowitz, I. H., 1970, Ground-water resources in the Eastern Oswego River Basin, New York: State of New York Conservation Department, Water Resources Commission Basin Planning Report ORB-2.

Kelly, T. E., 1974, Reconnaissance investigation of ground water in the Rio Grande drainage basin—with special emphasis on saline ground-water resources: Hydrologic Investigation Atlas HA-510, 4 sheets.

Kimball, B. A., 1987, Geochemical indicators used to determine source of saline water in Mesozoic aquifers, Montezuma Canyon area, Utah: U.S. Geological Survey Water-Supply Paper 2340,

Kimmel, G. E, 1963, Contamination of ground water by sea-water intrusion along Puget Sound, Washington, an area having abundant precipitation: U.S. Geological Survey Professional Paper 475-B, p. B182–B185.

Kister, L. R., Brown, S. G., Schumann, H. H., and Johnson, P. W., 1966, Maps showing fluoride content and salinity of ground water in the Wilcox Basin, Graham and Cochise Counties, Arizona: U.S. Geological Survey Hydrologic Investigations Atlas HA-214, scale 1:250,000, 2 sheets.

Knuth, Martin, Jackson, J. L., and Whittemore, D. O., 1990, An integrated approach to identifying the salinity source contaminating a ground-water supply: Groundwater, v. 28, no. 2, p. 207–214.

Kohout, F. A., 1960, Cyclic flow of salt water in the Biscayne Aquifer of Southeastern Florida: Journal of Geophysical Research, v. 65, no. 7, p. 2133–2141.

Kreitler, C. W., 1979, Nitrogen-isotope ratio studies of soils and groundwater nitrate from alluvial fan aquifers in Texas: Journal of Hydrology, v. 42, p. 147–170.

Kreitler, C. W., 1989, Hydrogeology of sedimentary basins: Journal of Hydrology, v. 106, p. 29–53.

Kreitler, C. W., and Jones, D. C., 1975, Natural soil nitrate—The cause of the nitrate contamination of ground water in Runnels County, Texas: Ground Water, v. 13, no. 1, p. 53–61.

Kreitler, C. W., Ragone, S. E., and Katz, G. B., 1978, N^{15}/N^{14} Ratios of ground-water nitrate, Long Island, New York: Ground Water, v. 16, no. 6, p. 404–409.

Kreitler, C. W., Chapman, J. B., and Knauth, L. P., 1984, Chemical and isotopical composition of waters from the Salina Ometepec, Baja California: The University of Texas at Austin, Bureau of Economic Geology Open-File Report OF-WTWI-1984-41, 13 p.

Kreitler, C. W., Akhter, M. S., Donnelly, A. C. A., and Wood, W. T., 1988, Hydrogeology of formations used for deep-well injection, Texas Gulf Coast: The University of Texas at Austin, Bureau of Economic Geology, report prepared for the U.S. Environmental Protection Agency under Cooperative Agreement ID No. CR812786-01-0, 204 p.

Kreitler, C. W., Akhter, M. S., and Donnelly, A. C. A, 1990, Hydrologic-hydrochemical characterization of Texas Frio Formation that is used for deep-well injection of chemical wastes: Journal of Environmental Geology and Water Sciences, v. 16, no. 2, p. 107–120.

Krieger, R. A., Hatchett, J. L., and Poole, J. L., 1957, Preliminary survey of the saline-water resources of the United States: U.S. Geological Survey Water-Supply Paper 1374, 172 p.

Krieger, R. A., and Hendrickson, G. E., 1960, Effects of Greensburg oil-field brines on the streams, wells, and springs of the Upper Green River: Kentucky Geological Survey, Series X, Report of Investigations No. 2, 36 p.

Kuhlmann, Mark, 1968, A study of chloride contamination in the Olentangy River in the vicinity of Columbus, Ohio: Unpublished Senior Thesis, Ohio State University, 26 p.

Kunkle, S. H. 1971, Effect of road salt on a Vermont Stream: Proceedings: Street Salting—Urban Water Quality Workshop, S.U.N.Y.: College of Forestry, Syracuse University, p. 48–61

Lamb, C. E., and Woodward, R., 1988, California ground-water quality, in Moody, D. W., Carr, Jerry, Chase, E. B., and Paulson, R. W., compilers, National Water Summary 1986—Hydrologic events and groundwater quality: U.S. Geological Survey Water-Supply Paper 2325, p. 173–180.

Land, L. S., 1987, The major ion chemistry of saline brines in sedimentary basins, in Banavar, J. R., Koplik, J., and Winkler, K. W., eds., Physics and Chemistry of Porous Media II: Ridgefield, Connecticut, American Institute of Physics Conference Proceedings 154, p. 160–179.

Land, L. S., and Prezbindowski, D. R., 1981, The origin and evolution of saline formation water, Lower Cretaceous Carbonates, South-Central Texas: Journal of Hydrology, v. 54, p. 51–74.

Land, L. S., and Prezbindowski, D. R., 1985, Chemical constraints and origins of four groups of Gulf Coast reservoir fluids: American Association of Petroleum Geologists Bulletin, v. 69, p. 119–126.

Larson, J. D., 1981, Distribution of salt water in the coastal plain aquifers of Virginia: U.S. Geological Survey Open-File Report 81-1013, 25 p.

Law, J. P., Jr., Davidson, J. M., and Reed, L. W., 1970, Degradation of water quality in irrigation return flows: Oklahoma Agricultural Experiment Station, Bulletin 684, p. 3–26.

Lehr, J. H., 1969, A study of ground-water contamination due to saline water disposal in the Morrow County oil field: Research Project Completion Report Project No. A-004-OHIO, submitted to Office of Water Resources Research, U.S. Department of the Interior, 81 p.

Leonard, R. B., 1964, A method for evaluating oil-field-brine pollution of the Walnut River in Kansas: U.S. Geological Survey Professional Paper 501-B, p. B173–B176.

Leonard, A. R., and Ward, P. E., 1962, Use of Na/Cl ratios to distinguish oil-field from salt-spring brines in Western Oklahoma: U.S. Geological Survey Professional Paper 450-B, p. B126–B127.

Leonard, R. B., and Kleinschmidt, M. K., 1976, Saline water in the Little Arkansas River Basin area, South-Central Kansas: The University of Kansas at Lawrence, Kansas Geological Survey Chemical Quality Series No. 3, 24 p.

Levings, G. W., 1984, Reconnaissance evaluation of contamination in the alluvial aquifer in the East Poplar oil field, Roosevelt County, Montana: U.S. Geological Survey Water-Resources Investigations Report 84-4174, 29 p.

Lico, M. S., Kharaka, Y. K., Carothers, W. W., and Wright, V. A., 1982, Methods for collection and analysis of geopressured geothermal and oil field waters: United States Geological Survey, Geological Survey Water-Supply Paper 2194, 21 p.

Lloyd, J. W., and Heathcote, J. A., 1985, Natural inorganic hydrochemistry in relation to ground water: Clarendon Press, Oxford, 294 p.

Lloyd, J. W., and Howard, K. W. F., 1979, Environmental isotope studies related to ground-water flow and saline encroachment in the Chalk Aquifer of Lincolnshire, England, in Isotope Hydrology 1978, International Symposium on Isotope Hydrology: International Atomic Energy Agency, Vienna, p. 311–325.

Lloyd, J. W., Howard, K. W. F., Pacey, N. R., and Tellam, J. H., 1982, The value of iodide as a parameter in the chemical characterization of ground waters: Journal of Hydrology, v. 57, p. 247–265.

Love, S. K., 1944, Cation-Exchange in Ground Water Contaminated with Sea Water near Miami, Florida: American Geophysical Union, Transactions, v. 25, p. 951–955.

Ludwig, A. H., 1972, Water resources of Hempstead, Lafayette, Little River, Miller, and Nevada Counties, Arkansas: U.S. Geological Survey Water Supply Paper 1998.

Lundegard, P. D., and Land, L. S., 1986, Carbon Dioxide and organic acids—Their role in porosity enhancement and cementation, Paleogene of the Texas Gulf Coast, in Gautier, D. L., ed., Roles of organic matter in sediment diagenesis: Society of Economic Paleontologists and Mineralogists Special Publication 38, p. 129–146.

Lusczynski, N. J., and Swarzenski, W. V., 1966, Salt-Water Encroachment in Southern Nassau and Southeastern Queens Counties, Long Island, New York: U.S. Geological Survey Water-Supply Paper 1613-F, 76 p.

Macumber, P. G., 1984, Hydrochemical processes in the regional ground-water discharge zones of the Murray Basin, Southeastern Australia, in Hitchon, Brian, and Wallick, E. I., eds., Proceedings First Canadian/American Conference on Hydrogeology—practical applications of ground-water geochemistry: National Water Well Association, Worthington, Ohio, p. 47–63.

Magaritz, Mordeckai, Nadler, Arie, Koyumdjisky, Hanna, and Dan, Joel, 1981, The use of Na/Cl ratios to trace solute sources in a semiarid zone: Water Resources Research, v. 17, no. 3, p. 602–608.

Magaritz, Mordeckai, and Luzier, J. E., 1985, Water–rock interactions and seawater–freshwater mixing effects in the coastal dune aquifer, Coos Bay, Oregon: Geochimica et Cosmochimica Acta, v. 49, p. 2515–2525.

Manheim, F. T., and Paull, C. K., 1981, Patterns of ground-water salinity changes in a deep continental-oceanic transect off the southeastern Atlantic Coast of the U.S.A., in Back, William, ed., Symposium on geochemistry of ground water: Journal of Hydrology, v. 54, nos. 1–3, p. 95–105.

Manheim, F. T., 1990, Potential saltwater intrusion into freshwater aquifers during exploration for and mining of phosphorite off the Southeastern United States: U.S. Geological Survey Circular 1033, p. 161–162.

Martin, Peter, 1982, Ground-water monitoring at Santa Barbara, California; Phase 2—effects of pumping on water levels and water quality in the Santa Barbara ground-water basin: U.S. Geological Survey Open-File Report No. 82-366, 53 p.

Mast, V. A., 1982, The use of ionic mixing curves in differentiating oil-field brine in a fresh-water aquifer: presented at the American Water Resources Association, The University of Texas at Austin, December 3, 1982, 16 p.

McGrain, Preston, and Thomas, G. R., 1951, Preliminary report on the natural brines of eastern Kentucky: Kentucky Geological Survey, Series IX, Report of Investigations No. 3, 22 p.

McGuire, M. J., Krasner, S. W., and Gramith, J. T., 1989, Comments on bromide levels in state project water and impacts on control of disinfection by-products: Metropolitan Water District of Southern California, 17 p.

McMillion, L. G., 1965, Hydrologic aspects of disposal of oil-field brines in Texas: Ground Water, v. 3, no. 4, p. 36–42.

McMillion, L. G., 1970, Ground-water reclamation by selective pumping: American Society of Mining Engineers Reprint No. 70-AG-55.

Meents, W. F., Bell, A. H., Rees, O. W. and Tilbury, W. G., 1952, Illinois oil-field brines—their geologic occurrence and chemical composition: Illinois State Geological Survey, Illinois Petroleum No. 66, 38 p.

Meisler, Harold, 1989, The occurrence and geochemistry of salty ground water in the Northern Atlantic Coastal Plain: U.S. Geological Survey Professional Paper 1404-D, 51 p.

Meisler, Harold, Leahy, P. P., and Knobel, L. L., 1985, Effect of eustatic sea-level changes on saltwater–freshwater relations in the Northern Atlantic Coastal Plain: U.S. Geological Survey Water-Supply Paper 2255, 28 p.

Meyer, F. W., 1971, Saline artesian water as a supplement: Journal of the American Works Association, v. 63, no. 2, p. 65–71.

Meyer, Michael, 1986, A summary of ground-water pollution problems in South Dakota: South Dakota Department of Water and Natural Resources, Office of Water Quality, 18 p.

Michie & Associates, 1988, Oil and gas industry water injection well corrosion: report prepared for the American Petroleum Institute, 25 p.

Michigan Department of Natural Resources, 1982, Assessment of groundwater contamination—inventory of sites: Lansing, 242 p.

Miller, D. W., ed., 1980, Waste Disposal Effects on Ground Water: Premier Press, 512 p.

Miller, D. W., DeLuca, F. A., and Tessier, T. L., 1974, Ground-water contamination in the northeast states: U.S. Environmental Protection Agency, Office of Research and Development, Robert S. Kerr Environmental Research Laboratory, Report EPA-660/2-74-056, 300 p.

Miller, J. C., Hackenberry, P. S., and DeLuca, F. A., 1977, Ground-water pollution problems in the Southeastern United States: U.S. Environmental Protection Agency, Office of Research and Development, Robert S. Kerr Environmental Research Laboratory, Ecological Research Series EPA-600/3-77-012, 329 p.

Miller, M. R., and others, 1980, Regional assessment of the saline-seep problem and a water-quality inventory of the Montana plain: Montana Bureau of Mines and Geology, Open-File Report 42, 414 p.

Mink, J. F., 1960, Some geochemical aspects of sea-water intrusion in a island aquifer: General Assembly of Helsinki, International Association of Scientific Hydrology, Publication 52, p. 424–439.

Monterey County Flood Control & Water Conservation District, 1989, Sources of saline intrusion in the 400-foot aquifer, Castroville area, California: Report prepared by D. K. Todd Consulting Engineers, Inc., Berkeley, California, 43 p.

Mook, W. G., 1970, Stable carbon and oxygen isotopes of natural waters in the Netherlands, *in* Isotope Hydrology: International Atomic Energy Agency, Vienna, p. 163–190.

Moore, J. W., and Welch, R. C., 1977, Environmental aspects, brine usage for highway purposes: final report conducted for the Arkansas State Highway Department, Highway Research Project 44, 133 p.

Morris, E. E., 1988, Arkansas ground-water quality, *in* Moody, D. W., Carr, Jerry, Chase, E. B., and Paulson, R. W., compilers, National Water Summary 1986—Hydrologic events and groundwater quality: U.S. Geological Survey Water-Supply Paper 2325, p. 165–172.

Morris, E. E. and Bush, W. V., 1986, Extent and source of saltwater intrusion into the alluvial aquifer near Brinkley, Arkansas, 1984: Arkansas Geological Commission, Water Resources Circular 15, 35 p.

Morrissey, D. J., and Regan, J. M., 1988, New Hampshire ground-water quality, *in* Moody, D. W., Carr, Jerry, Chase, E. B., and Paulson, R. W., compilers, National Water Summary 1986—Hydrologic events and groundwater quality: U.S. Geological Survey Water-Supply Paper 2325, p. 363–368.

Morton, R. A., and Land, L. S., 1987, Regional variations in formation water chemistry, Frio Formation (Oligocene), Texas Gulf Coast: American Association of Petroleum Geologists Bulletin, v. 71, no. 2, p. 191–206.

Morton, R. B., 1986, Effects of brine on the chemical quality of water in parts of Creek, Lincoln, Okfuskee, Payne, Pottawatomie, and Seminole Counties, Oklahoma: Oklahoma Geological Survey Circular 89, 37 p.

Muller, A. B., 1974, Interdisciplinary modeling in the analysis of the salinity problems of the Safford Valley: Water Resources Bulletin, v. 10, no. 2, p. 245–255.

Mullican, W. F., 1988, Subsidence and collapse at Texas salt domes: The University of Texas at Austin, Bureau of Economic Geology, Geological Circular 88-2, 35 p.

Murphy, E. C., and Kehew, A. E., 1984. The effect of oil and gas well drilling fluids on shallow groundwater in western North Dakota. Report of Investigations No. 82, North Dakota Geological Survey, 156 p.

Nadler, A., Magaritz, M., and Mazar, E., 1981, Chemical reactions of seawater with rocks and freshwater—experimental and field observations on brackish waters in Israel: Geochimica et Cosmochimica Acta, v. 44, p. 879–886.

National Resources and Agriculture Committee, 1973, Interim report of the special commission on salt contamination of water supplies and related matters: Massachusetts Senate Document 1485, January 1973.

Nativ, Ronit, 1988, Hydrogeology and hydrochemistry of the Ogallala aquifer, Southern High Plains, Texas Panhandle and Eastern New Mexico: The University of Texas at Austin, Bureau of Economic Geology Report of Investigations 177, 64 p.

Neffendorf, D. W., 1978, Statewide saline seep survey of Texas: Master's Thesis, Texas A&M University, 67 p.

Newport, B. D., 1977, Salt-water intrusion in the United States: U.S. Environmental Protection Agency, Office of Research and Development, Robert S. Kerr Environmental Research Laboratory Report EPA-600/8-77-011, 30 p.

Nordstrom, D. K., 1983, Preliminary data on the geochemical characteristics of groundwater at Stripa: Proceedings NEA Workshop on Geological Disposal of Radioactive Waste—in situ experiments in granite, Nuclear Energy Agency, p. 143–153.

Novak, S. A., and Eckstein, Yoram, 1988, Hydrochemical characterization of brines and identification of brine contamination in aquifers: Ground Water, v. 26, no. 3, p. 317–324.

Nyman, D. J., 1978, The occurrence of saline ground water in the coastal area of southwestern Louisiana: EOS, American Geophysical Union, Transactions, v. 59, no. 12, p. 1065.

O'Hare, Margaret, Curry, D. S., Atkinson, S. F., Lee, S. B., and Canter, L. W., 1985, Contamination of ground water in the contiguous United States from usage of agricultural chemicals (pesticides and fertilizers): Environmental and Ground Water Institute, Norman, University of Oklahoma, variously paginated.

Oklahoma Water Resources Board, 1975, Salt water detection in the Cimarron Terrace, Oklahoma: U.S. Environmental Protection Agency, Office of Research and Development, Robert S. Kerr Environmental Research Laboratory, Report EPA-660/3-74-033, 100 p.

Ong, Kim, 1988, New Mexico ground-water quality, in Moody, D. W., Carr, Jerry, Chase, E. B., and Paulson, R. W., compilers, National Water Summary 1986—Hydrologic events and groundwater quality: U.S. Geological Survey Water-Supply Paper 2325, p. 377–384.

Orr, B. R., and Myers, R. G., 1986, Water resources in basin-fill deposits in the Tularosa Basin, New Mexico: U.S. Geological Survey, Water–Resources Investigations Report 85-4219, 94 p.

Parker, J. W., 1969, Water history of Cretaceous aquifers, East Texas Basin: Chemical Geology, v. 4, nos. 1–2, p. 111–133.

Parliman, D. J., 1988, Idaho ground-water quality, in Moody, D. W., Carr, Jerry, Chase, E. B., and Paulson, R. W., compilers, National Water Summary 1986—Hydrologic events and groundwater quality: U.S. Geological Survey Water-Supply Paper 2325, p. 229–236.

Patterson, R. J., and Kinsman, D. J. J., 1975, Marine and continental ground-water sources in a Persian Gulf coastal sabkha, in Frost, S. H., Weiss, M. P., and Saunders, J. B., Reefs and related carbonates—ecology and sedimentology: American Association of Petroleum Geologists, Studies in Geology No. 4, p. 381–397.

Peck, A. J., 1978, Salinization of non-irrigated soils and associated streams—a review: Australian Journal of Soil Research, v. 16, p. 157–168.

Pennino, J. D., 1988, There's no such thing as a representative ground water sample: Ground Water Monitoring Review, v. 8, no. 3, p. 4–9.

PennWell Publishing Co., 1982, Oil and gas fields in the United States: PennWell Publishing Company, Tulsa, Oklahoma, map scale 1:3,530,000.

Pettyjohn, W. A., 1971, Water Pollution by Oil-Field Brines and Related Industrial Wastes in Ohio: Ohio Journal of Science, v. 71, no. 5, p. 257–269.

Pettyjohn, W. A., 1982, Cause and effect of cyclic changes in ground-water quality: Ground Water Monitoring Review, Winter 1982, p. 43–49.

Pierce, W. G., and Rich, E. I., 1962, Summary of rock salt deposits in the United States as possible storage sites for radioactive waste materials: Geological Survey Bulletin 1148, prepared on behalf of the U.S. Atomic Energy Commission, 91 p.

Piper, A. M., 1944, A graphic procedure in the geochemical interpretation of water analyses: American Geophysical Union, Transactions, v. 25, p. 914–923.

Piper, A. M., Garrett, A. A., and others, 1953, Native and contaminated ground waters in the Long Beach–Santa Ana area, California: U. S. Geological Survey, Water-Supply Paper 1136, 320 p.

Poland, J. F., Garrett, A. A., and Sinnott, A., 1959, Geology, hydrology, and chemical character of ground waters in the Torrance–Santa Monica area, California: U. S. Geological Survey, Water-Supply Paper 1461, 425 p.

Pomper, A. B., 1981, Hydrochemical observations in the subsoil of the western part of the Netherlands: Proceedings, Seventh salt water intrusion meeting, Uppsala, Sweden, p. 101–111.

Poth, C. W., 1962, The occurrence of brine in western Pennsylvania: Pennsylvania Geological Survey, Fourth Series, Bulletin M 47, 53 p.

Price, R. D., 1979, Occurrence, quality, and quantity of ground water in Wilbarger County, Texas: Texas Department of Water Resources Report 240, 229 p.

Price, R. M., 1988, Geochemical investigation of salt-water intrusion along the coast of Mallorca, Spain: University of Virginia, Charlottesville, Master's thesis, 186 p.

Puls, R. W., Eychaner, J. H., and Powell, R. M., 1990, Colloidal-facilitated transport of inorganic contaminants in ground water, Part I. Sampling considerations: U.S. Environmental Protection Agency, Robert S. Kerr Environmental Research Laboratory, EPA/600/M-90/023, 12 p.

Randall, A.D., 1972, Records of Wells and Test Borings in Susquehanna River Basin, New York: New York State Department of Environmental Conservation Bulletin No 69.

Rau, J. L., 1973, Effects of brining and salt by-products operations on the surface and ground-water resources of the Muskingum Basin, Ohio, in Coogan, A. H. ed., Fourth Symposium on Salt: Northern Ohio Geological Society, v. 1, p. 369–386.

Reeves, C. C., Jr., 1970, Saline lake basins of the Southern High Plains, in Mattox, R. B., Ground-water salinity: Symposium, 46th Annual Meeting of the Southwestern and Rocky Mountain Division of the

American Association of the Advancement of Science, April 23–24, Las Vegas, New Mexico, p. 64–69.

Reichenbaugh, R. C., 1972, Sea-water intrusion in the upper part of the Floridan aquifer in coastal Pasco County, Florida, 1969: U.S. Geological Survey, Florida Bureau of Geology Map Series no. 47.

Rice, G. F., Muller, D. L., and Brinkman, J. E., 1988, Reliability of chemical analyses of water samples—the experience of the UMTRA project: Ground Water Monitoring Review, v. 8, no. 3, p. 71-75.

Richter, B. C., and Kreitler, C. W., 1986a, Geochemistry of salt water beneath the Rolling Plains, North-Central Texas: Ground Water, v. 24, no. 6, p. 735–742.

Richter, B. C., and Kreitler, C. W., 1986b, Geochemistry of salt-spring and shallow subsurface brines in the Rolling Plains of Texas and southwestern Oklahoma: The University of Texas at Austin, Bureau of Economic Geology Report of Investigations No. 155, 47 p.

Richter, B. C., Dutton, A. R., and Kreitler, C. W., 1990, Identification of sources and mechanisms of salt-water pollution affecting ground-water quality—a case study, West Texas: The University of Texas at Austin, Bureau of Economic Geology Report of Investigations No. 191, 43 p.

Ritter, W. F., and Chirnside, A. E. M., 1982, Ground-water quality in selected areas of Kent and Sussex Counties, Delaware: Newark, University of Delaware Agricultural Engineering Department, 229 p.

Rittenhouse, Gordon, 1967, Bromine in oil-field waters and its use in determining possibilities of origin of these waters: The American Association of Petroleum Geologists Bulletin, p. 2430–2440.

Rittenhouse, Gordon, Fulton, R. B., III, Grabowski, R. J., and Bernard, J. L., 1969, Minor elements in oil-field waters: Chemical Geology, v. 4, p. 189–209.

Roberson, C. E., Feth, J. H., Seaber, P. R., and Anderson, Peter, 1963, Differences between field and laboratory determinations of pH, alkalinity, and specific conductance of natural water, *in* Short papers in geology and hydrology: U.S. Geological Survey Professional Paper 475-C, p. C212–C217.

Robinove, C. J., Langford, R. H., and Brookhart, J. W., 1958, Saline-water resources of North Dakota: U.S. Geological Survey Water-Supply Paper 1428, 72 p.

Roedder, E., 1984, Fluid inclusions: Mineralogical Society of America, Rev. in Min. 12, p. 644.

Rold, J. W., 1971, Pollution problem in the "Oil Patch": American Association of Petroleum Geologists Bulletin v. 55, no. 6, p. 807–809.

Rumer, R. R., Jr., Apmann, R. P., and Chien, C. C., 1973, Runoff of deicing salt in Buffalo, New York, *in* Coogan, A. H. ed., Fourth Symposium on Salt: Northern Ohio Geological Society, v. 1., p. 407–411.

Sabol, G. V., Bouwer, H., Wierenga, P. J., 1987, Irrigation effects in Arizona and New Mexico: American Society of Civil Engineers, Journal of Irrigation and Drainage Engineering (ASCE), v. 113, no. 1, p. 30–48.

Saffigna, P. G., and Keeney, D. R., 1977, Nitrate and chloride in ground water under irrigated agriculture in central Wisconsin: Ground Water, v. 15, no. 2, p. 170–177.

Salt Institute, undated, Survey of salt, calcium chloride, and abrasive use in the United States and Canada for 81/82 and 82/83: Salt Institute, Alexandria, Virginia, 4 p., 9 tables.

Sandoval, F. M. and Benz, L. C., 1966, Effect of bare fallow, barley, and grass on salinity of a soil over a saline water table: Soil Science Society of America Proceedings, v. 30, p. 392–396.

Sayre, A. N., 1937, Geology and ground-water resources of Duval County, Texas: U.S. Geological Survey Water Supply Paper 776, 116 p.

Scalf, M. R., Keely, M. R., and LaFevers, C. J., 1973, Ground-water pollution in the south-central states: U.S. Environmental Protection Agency, Office of Research and Development, National Environmental Research Center, Environmental Protection Technology Series EPA-R2-73-268, 135 p.

Schaefer, F. L., 1983, Distribution of chloride concentrations in the principal aquifers of the New Jersey Coastal Plain, 1977–81: U.S. Geological Survey Water-Resources Investigations Report 83-4061, 56 p.

Schaefer, F. L., and Walker, R. L., 1981, Salt-water intrusion into the Old Bridge Aquifer in the Keyport-Union Beach area of Monmouth County, New Jersey: U.S. Geological Survey Water-Supply Paper 2184, 21 p.

Scheidt, M. E., 1967, Environmental effects of highways: Journal of the Sanitary Engineering Division, Proceedings of the American Society of Civil Engineers, v. 93, no. SA5, paper no. 5509.

Schmidt, Ludwig, and Devine, J. M., 1929, The disposal of oil-field brines: U.S. Department of the Interior, Bureau of Mines, Report of Investigations 2945, 17 p.

Schmidt, V., and McDonald, D. A., 1979, The role of secondary porosity in the course of sandstone diagenesis, Aspects of Diagenesis, *in* Scholle, P. A., and Schluger, R., Society of Economic Paleontologists and Mineralogists Special Publication 26, p. 175–208.

Schoeller, Henri, 1935, Utilité de la notion des exchanges de bases pour la comparison des eaux souterraines: France, Société Géologie Comptes rendus Sommaire et Bulletin Série 5, v. 5, p. 651–657.

Schofield, J. C., 1956, Methods of distinguishing sea–ground-water from hydrothermal water: New Zealand Journal of Science and Technology, General Research Section, v. 37, no. 5, p. 597–602.

Schraufnagel, F. H., 1967, Pollution Aspect associated with chemical de-icing: Highway Research Record Report 193, p. 22–23.

Schwartz, F. W., 1974, The origin of chemical variations in ground waters from a small watershed in southwestern Ontario: Canadian Journal Earth Science, v. 11, no. 7, p. 893–904.

Senger, R. K., Collins, E. W., and Kreitler, C. W., 1990, Hydrogeology of the northern segment of the Edwards aquifer, Austin region: The University of Texas at Austin, Bureau of Economic Geology, Report of Investigations No. 192, 58 p.

Shamburger, V. M., Jr., 1958, Alleged well contamination in relation to brine disposal in Clemville, Matagorda County: Texas Board of Water Engineers Contamination Report, 10 p.

Shedlock, R. J., 1978, Saline water in the glacial outwash aquifer near Vincennes, Indiana: EOS, American Geophysical Union, Transactions, v. 59, no. 12, p. 1065.

Shirazi, G. A., Allen, D. L., and Madden, M. P., 1976, An investigation of brine contamination of soils and ground water from a slush pit near burns flat, Oklahoma: Oklahoma Water Resources Board, 91 p.

Shows, T. A., Broussard, W. L., and Humphreys, C. P., Jr., 1966, Water for industrial development in Forrest, Green, Jones, Perry and Wayne Counties, Mississippi: Mississippi Research and Development Center.

Sidenvall, Jan, 1981, Fossil ground water of marine origin in the Uppsala area, Sweden: Proceedings, Seventh Salt Water Intrusion meeting, Uppsala, Sweden, p. 40–44.

Siebenthal, C. E., 1910, Geology and water resources of the San Luis Valley, Colorado: U.S Geological Survey Water-Supply Paper 240, 128 p.

Siple, G. E., 1969, Salt-water encroachment of tertiary limestones along coastal South Carolina: South Carolina Development Board, Division of Geology, Geologic Notes 13, v. 2, p. 51–65.

Skibitzke, H. E., Bennett, R. R., DaCosta, J. A., Lewis, D. D., and Maddock, T., Jr., 1961, Symposium on history of development of water supply in an arid area in southwestern United States, Salt River Valley, Arizona: International Association of Scientific Hydrology, Publication 57, p. 706–742.

Skogerboe, G. V., and Law, J. P., 1971, Research needs for irrigation return flow quality control: U.S. Environmental Protection Agency, Water Pollution Control Research Series 13030.

Skogerboe, G. V., and Walker, W. R., 1973, Salt pickup from agricultural lands in the Grand Valley of Colorado: Journal of Environmental Quality, v. 2, no. 3, p. 377–382.

Slack, L. J., and Planert, Michael, 1988, Alabama ground-water quality, in, Moody, D. W., Carr, Jerry, Chase, E. B., and Paulson, R. W., compilers, National Water Summary 1986—Hydrologic events and groundwater quality: U.S. Geological Survey Water-Supply Paper 2325, p. 143–148.

Smith, J. F., 1966, Imperial Valley salt balance: Public Information Office, Imperial Irrigation District, El Centro, California.

Smith, Z. A., 1989, Groundwater in the West: Academic Press, San Diego, California, 308 p.

Snow, M. S., Kahl, J. S., Norton, S. A., and Olson, Christine, 1990, Geochemical determination of salinity sources in ground water wells in Maine: Proceedings of the 1990 FOCUS conference on eastern regional ground water issues, Ground Water Management Book 3, National Water Well Association, p. 313–327.

Snyder, R. H., Buehrer, L. C., and Bell, F. O., 1972, A survey of the subsurface saline water of Texas, v. 1: Texas Water Development Board Report 157, 113 p.

Sorensen, H. O., and Segall, R. T., 1973, Natural brines of the Detroit River Group, Michigan Basin, in Coogan, A. H. ed., Fourth Symposium on Salt: Northern Ohio Geological Society, v. 1, p. 91–99.

252

Sowayan, A. M., and Allayla, Rashid, 1989, Origin of the saline ground water in Wadi Ar-Rumah, Saudi Arabia: Ground Water, v. 27, no. 4, p. 481–490.

Speiran, G. K., Oldham, R. W., Duncan, D., and Knox, R. L., 1988, South Carolina ground-water quality, *in* Moody, D. W., Carr, Jerry, Chase, E. B., and Paulson, R. W., compilers, National Water Summary 1986—Hydrologic events and groundwater quality: U.S. Geological Survey Water-Supply Paper 2325, p. 449–456.

Sproul, C. R., Boggess, D. H., and Woodard, H. J., 1972, Saline-water intrusion from deep artesian sources in the McGregor Isles area of Lee County, Florida: State of Florida, Department of Natural Resources, Division of Interior Resources, Bureau of Geology Information Circular No. 75, 30 p.

Spruill, T. B., 1985, Statistical evaluation of effects of irrigation on chemical quality of ground water and base flow in three river valleys in north-central Kansas: U.S. Geological Survey Water-Resources Investigations Report 85-4158, 64 p.

Stafford, M. R., 1987, Hydrogeology, ground-water chemistry, and resistivity of a contaminated shallow aquifer system in southern Bond County, Illinois: Southern Illinois University, Master's thesis.

Steinkampf, W. C., 1982, Origins and distribution of saline ground waters in the Floridan aquifer in coastal southwest Florida: U.S. Geological Survey Water-Resources Investigations Report 82-4052, 34 p.

Stephens, J. C., 1974, Hydrologic Reconnaissance of the northern Great Salt Lake Desert and summary hydrologic reconnaissance of northwestern Utah: Utah Department of Natural Resources Technical Publication No. 42, 55 p.

Stringfield, V. T., and LeGrand, H. E., 1969, Relation of sea water to fresh water in carbonate rocks in coastal areas, with special reference to Florida, U.S.A., and Cephalonia (Kephallinia), Greece: Journal of Hydrology 9, p. 387–404.

Subra, W. A., 1990, Unsuccessful oil field waste disposal techniques in Vermilion Parish, Louisiana, *in* Proceedings of the First International Symposium on Oil and Gas Exploration and Production Waste Management Practices, New Orleans, Louisiana: U.S. Environmental Protection Agency, p. 995–999.

Sundstrom, R. W., Hastings, W. W., and Broadhurts, W. L., 1948, Public water supply in eastern Texas: U.S. Geological Survey Water-Supply Paper 1047, 285 p.

Swarzenski, W. V., 1963. Hydrogeology of Northwestern Nassau and Northeastern Queens Counties: Long Island, New York, U.S. Geological Survey Water-Supply Paper 1657.

Sweeten, J. M., 1990, Cattle feedlot waste management practices for water and air pollution control: Texas Agricultural Extension Service, B-1671, 20 p.

Sweeten, J. M., Safley, L. M., and Melvin, S. W., 1981, Sludge removal from lagoons and holding ponds—case studies, *in* Livestock waste—a renewable resource: Proceedings of the Fourth International Symposium on Livestock Wastes, American Society of Agricultural Engineers, St. Joseph, Michigan, p. 204–210.

253

Sylvester, R. O., and Seabloom, R. W., 1963, Quality and significance of irrigation return flow: Journal of the Irrigation and Drainage Division, American Society of Civil Engineers, v. 89, no. IR3, p. 1–27.

Taylor, O. J., 1983, Missouri Basin Region, in Todd, D. K., ed., Ground-water resources of the United States: Premier Press, Berkeley, California, p. 297–343.

Templeton, E. E., and Associates, 1980, Environmentally acceptable disposal of salt brine produced with oil and gas: Ohio Water Development Authority, 52 p.

Texas State Soil and Water Conservation Board, 1984, Preliminary Report-Assessment of soil salinity in Texas: Texas Department of Water Resources, Contract No. 14-40019 1AC(84-85) 1038, Statewide Inventory of Salinity Problems.

Texas Water Commission, 1989, Ground-Water Quality of Texas—an overview of natural and man-affected conditions: Report 89-01, 197 p.

Thomas, J. M., and Hoffman, R. J., 1988, Nevada ground-water quality, in Moody, D. W., Carr, Jerry, Chase, E. B., and Paulson, R. W., compilers, National Water Summary 1986—Hydrologic events and groundwater quality: U.S. Geological Survey Water-Supply Paper 2325, p. 355–362.

Thompson, R. G., and Custer, S. G., 1976, Shallow ground-water salinization in dryland-farm areas of Montana: Montana Universities Joint Water Resources Research Center, MUJWRRC Report No. 79, 214 p.

Thorne, W., and Peterson, H. B., 1967, Salinity in United States waters, in Brady, N. C., ed., Agriculture and the quality of our environment: American Association for the Advancement of Science, Washington, D.C., Publication 85, p. 221–240.

Tremblay, J. J., 1973, Salt Water Intrusion in the Summerside Area, P.E.I.: Ground Water, v. 11, No. 2, p. 21–27.

Truesdell, A. H., 1976, Summary of section III, geochemical techniques in exploration, in Proceedings, Second United Nations Symposium on the Development and Use of Geothermal Resources, San Francisco, California, 20–29 May, 1975 Berkeley, California, Lawrence Berkeley Laboratory, v. 1.

Tsunogai, S., 1975, Sea salt particles transported to the land: Tellus, v. 27, no. 1, p. 51–58.

Turner, S. F., and Foster, M. D., 1934, A Study of Salt-Water Encroachment in the Galveston Area, Texas, American Geophysical Union Transactions, v. 15, p. 432–435.

Twenter, F. R., and Cummings, T. R., 1985, Quality of ground water in Monitor and Williams Townships, Bay County, Michigan: U.S. Geological Survey Water-Resources Investigations Report 85-4110, 41 p.

UNESCO, 1980, Aquifer contamination and protection: Project 8.3 of the International Hydrologic Programme, prepared by the Project Working Group, Jackson, R. E., chairman and general editor, 119 p.

Unger, S. G., 1977, Environmental implications of trends in agriculture and silviculture: U.S. Environmental Protection Agency, Athens, Georgia, EPA-600/3-77-121, 188 p.

Upson, J. E., 1966, Relationships of fresh and salty ground water in the northern Atlantic Coastal Plain of the United States: U.S. Geological Survey Research, U.S Geological Survey Professional Paper 550-C, p. c235–c243.

Urish, D. W., and Ozbilgin, M. M., 1985, Mean sea level—an elusive boundary, in The second annual eastern regional ground water conference, July 16–18, 1985, Portland, Maine, p. 348–358.

Utah State University Foundation, 1969, Characteristics and pollution problems of irrigation return flow: Report prepared for U.S. Department of the Interior, Federal Water Pollution Control Administration, Robert S. Kerr Water Research Center, Ada, Oklahoma, under Contract Number 14-12-408, 236 p.

U.S. Department of Agriculture, 1975, Erosion, sediment, and related salt problems and treatment opportunities: Soil Conservation Service, Special Projects Division, Golden, Colorado.

U.S. Department of Agriculture, 1983, Dryland salinity study—Texas-Cooperative River Basin Survey, prepared by United States Department of Agriculture Soil Conservation Service Economic Research Service in cooperation with Texas Department of Water Resources-Texas State Soil and Water Conservation Board.

U.S. Environmental Protection Agency, 1973, Identification and control of pollution from salt water intrusion: EPA Office of Air and Water Programs, Water Quality and Non-Point Source Control Division, Report EPA-430/9-73-013, 94 p.

U.S. Environmental Protection Agency, 1975, Water programs—national interim primary drinking water regulations: Federal Register, v. 40, no. 248.

U.S. Environmental Protection Agency, 1977, National secondary drinking water regulations: Federal Register, v. 42, no. 62, pt. 1, p. 17143–17147.

U.S. Environmental Protection Agency, 1989, Proposed requirements for 38 inorganic and organic drinking-water contaminants: Office of Drinking Water, Washington, D.C. 20460, 10 p.

U.S. Geological Survey, 1984, National water summary 1983—Hydrologic Events and Issues, United States Geological Survey Water-Supply Paper 2250: U.S. Geological Survey, 243 p.

U.S. Salinity Laboratory, 1954, Diagnosis and improvement of saline and alkaline soils: U.S. Department of Agriculture Handbook 60, 160 p.

U.S. Water News, 1990a, Bromide doesn't really help California's delta headaches: v. 6, no. 7.

U.S. Water News, 1990b, Salt is invading Florida Gulf Coast: v. 7, no. 3, p. 12.

Van Biersel, T. P., 1985, Hydrogeology and chemistry of an oil-field brine plume within a shallow aquifer system in southern Bond County, Illinois: Southern Illinois University, Master's thesis.

Van Denburgh, A. S., and Rush, F. E., 1974, Water-resources appraisal of Railroad and Penoyer Valleys, east-central Nevada: Nevada Division of Water Resources, Water Resources Reconnaissance Series, Report no. 60, 61 p.

van der Leeden, Frits, Cerrillo, L. A., and Miller, D. W., 1975, Ground-water pollution problems in the Northwestern United States: U.S. Environmental Protection Agency, Office of Research and

Development, Robert S. Kerr Environmental Research Laboratory, Report EPA-660/3-75-018, 341 p.

van Everdingen, R. O., 1971, Surface-water composition in southern Manitoba reflecting discharge of saline subsurface waters and subsurface solution of evaporites, in, Turnock, A. C., ed., Geoscience Studies in Manitoba: Geological Association of Canada, Special Paper No. 9, p. 343–352.

Van Sickle, Virginia, and Groat, C. G., 1990, Oil field brines—another problem for Louisiana's coastal wetlands, in, Proceedings of the First International Symposium on Oil and Gas Exploration and Production Waste Management Practices, New Orleans, Louisiana: U.S. Environmental Protection Agency, p. 659–675.

Voelker, D. C., and Clarke, R. P., 1988, Illinois ground-water quality, in Moody, D. W., Carr, Jerry, Chase, E. B., and Paulson, R. W., compilers, National Water Summary 1986—Hydrologic events and groundwater quality: U.S. Geological Survey Water-Supply Paper 2325, p. 237–244.

Vovk, I. F., 1981, Radiolytic model of formation of brine compositions in the crystalline basement of Shields: Geochemistry International, p. 80–93.

Waddell, K. M., and Maxell, M. H., 1988, Utah ground-water quality, in Moody, D. W., Carr, Jerry, Chase, E. B., and Paulson, R. W., compilers, National Water Summary 1986—Hydrologic events and groundwater quality: U.S. Geological Survey Water-Supply Paper 2325, p. 493–500.

Wait, R. L., and Gregg, D. O., 1973, Hydrology and chloride contamination of the principal artesian aquifer in Glynn County, Georgia: Water Resources Survey of Georgia, Earth and Water Division, Hydrologic Report 1.

Wait, R. L., and McCollum, M. J., 1963, Contamination of fresh-water aquifers through an unplugged oil-test well in Glynn County, Georgia: Georgia Geological Survey Mineral Newsletter, v. 16, nos. 3–4, p. 74–80.

Walker, W. H., 1970, Salt piling—a source of water supply pollution: Pollution Engineer, v. 1, no. 4, p. 30–33.

Waller, B. G., and Howie, Barbara, 1988, Determining nonpoint-source contamination by agricultural chemicals in an unconfined aquifer, Dade County, Florida: Procedures and preliminary results, in Ground-water contamination: Field Methods, ASTM STP 963, American Society for Testing and Materials, Philadelphia, p. 459–467.

Wallick, E. I., Krouse, H. R., and Shakur, Asif, 1984, Environmental isotopes—principles and applications in ground water geochemical studies in Alberta, Canada, in Hitchon, Brian, and Wallick, E. I., eds., Proceedings, First Canadian/American Conference on Hydrogeology—Practical applications of ground-water geochemistry: National Water Well Association, Worthington, Ohio, p. 249–266.

Ward, P. E., 1961, Geology and groundwater features of salt springs, seeps, and plains in the Arkansas and Red River Basin of western Oklahoma and adjacent to Kansas and Texas: U.S. Geological Survey, Open-File Report No. 63-132, 94 p.

Welch, A. H., and Preissler, A. M., 1986, Aqueous geochemistry of the Bradys Hot Springs geothermal area, Churchill County, Nevada: U.S. Geological Survey Water-Supply Paper 2290, p. 17–36.

Wheeler J. C., and Maclin, L. B., 1988, Maryland and the District of Columbia ground-water quality, *in* Moody, D. W., Carr, Jerry, Chase, E. B., and Paulson, R. W., compilers, National Water Summary 1986—Hydrologic events and groundwater quality: U.S. Geological Survey Water-Supply Paper 2325, p. 287–296.

Whitehead, D. C. 1974, The influence of organic matter, chalk and sesquioxides on the solubility of iodine, elemental iodine and iodate incubated with soil: Journal Soil Science, v. 4, p. 461–470.

Whitehead, H. C., and Feth, J. H., 1961, Recent chemical analyses of waters from several closed-basin lakes and their tributaries in the western United States: Geological Society of America Bulletin, v. 72, p. 1421–1426.

Whiteman, C. D., Jr, 1979, Salt-water encroachment in the '600-foot' and '1,500-foot' sands of the Baton Rouge area, Louisiana, 1966–78, including a discussion of saltwater in other sands: Louisiana Department of Transportation and Development, Office of Public Works, Water Resources Technical Report No. 19, 49 p.

Whittemore, D. O., 1984, Geochemical identification of salinity sources, *in* French, R. H., Salinity in watercourses and reservoirs: Proceedings of the 1983 International Symposium on State-of-the-Art Control of Salinity, Salt Lake City, Utah, p. 505–514.

Whittemore, D. O., 1988, Bromide as a tracer in ground-water studies: Geochemistry and analytical determination: National Water Well Association, Proceedings, Ground Water Geochemistry Conference, Denver, Colorado, February 16–18, 1988, p. 339–359.

Whittemore, D. O., and Pollock, L. M., 1979, Determination of salinity sources in water resources of Kansas by minor alkali metal and halide chemistry: Kansas Water Resources Research Institute, Consultant's report to Office of Water Research and Technology, U.S. Department of the Interior, Washington, D.C., 37 p.

Wilcox, L. V., and Durum, W. H., 1967, Quality of irrigation water, *in* Hagan, R. M., Haise, H. R., and Edminster, T. W., eds., Irrigation of agriculture lands: American Society of Agronomy, Madison, no. 11, p. 104–122.

Wilder, H. B., Robinson, T. M., and Lindskov, K. L., 1978, Water resources of northeast North Carolina: U.S. Geological Survey, Water Resources Investigations 77-81, 113 p.

Williams, C. C. and C. K. Bayne, 1946, Ground-Water Conditions in Elm Creek Valley, Barber County, Kansas, with Special Reference to Contamination of Ground Water by Oil Field Brine: Kansas State Geological Survey Bulletin 64, 1946 Reports of Studies, part 3, p. 77–124.

Wilmoth, B. A., 1972, Salty ground water and meteoric flushing of contaminated aquifers in West Virginia: Ground Water, v. 10, no. 1, p. 99–105.

Wilson, W. E., 1982, Estimated effects of projected ground-water withdrawals on movement of the saltwater front in the Floridan aquifer, 1976–2000, West Central Florida: U.S. Geological Survey Water-Supply Paper 2189, 24 p.

Winslow, A.G., Doyle, W. W., and Wood, L. A., 1957, Salt Water and Its Relation to Fresh Ground Water in Harris County, Texas: U.S. Geological Survey Water-Supply Paper 1360-F, 33 p.

Woessner, W. W., Mifflin, M. D., French, R. H., Elzeftawy, Atef, and Zimmerman, D. E., 1984, Salinity balance in the Lower Virgin River Basin, Nevada and Arizona, *in* French, R. H., ed., Salinity in watercourses and reservoirs, Proceedings, 1983 International Symposium on State-of-the-Art Control of Salinity, Salt Lake City, Utah: Butterworth Publishers, Boston, p. 145–156.

Wood, W. W., 1976, Guidelines for collection and field analysis of ground-water samples for selected unstable constituents, 1976: Techniques of Water-Resources Investigations of the United States Geological Survey, Book 1, Chapter D2, 24 p.

Wood, W. W., and Jones, B. F., 1990, Origin of solutes in saline lakes and springs on the southern High Plains of Texas and New Mexico, *in* Gustavson, T. C., ed., Geologic framework and regional hydrology—Upper Cenozoic Blackwater Draw and Ogallala Formation, Great Plains: The University of Texas at Austin, Bureau of Economic Geology, p. 193–208.

Woodruff, K. D., 1969, The occurrence of saline ground water in Delaware aquifers: Delaware Geological Survey, Report of Investigations No. 13, 45 p.

Wyoming Department of Environmental Quality, 1986, Statewide water quality assessment 1986—Executive summary: Cheyenne, Wyoming, 26 p.

Yeh, H. W., 1980, D/H ratios and late-stage dehydration of shales during burial: Geochimica et Cosmochimica Acta, v. 44, p. 341–352.

Zack, Allen, and Roberts, Ivan, 1988, The geochemical evolution of aqueous sodium in the Black Creek Aquifer, Horry and Georgetown Counties, South Carolina: U.S. Geological Survey Water-Supply Paper 2324, 15 p.